INFECTION CONTROL
Dilemmas and Practical Solutions

INFECTION CONTROL
Dilemmas and Practical Solutions

Edited by

Kenneth R. Cundy
Temple University School of Medicine
Philadelphia, Pennsylvania

Bruce Kleger
Pennsylvania Department of Health
Lionville, Pennsylvania

Eileen Hinks
United Hospitals, Inc.
Philadelphia, Pennsylvania

and

Linda A. Miller
Holy Redeemer Hospital and Medical Center
Meadowbrook, Pennsylvania

PLENUM PRESS • NEW YORK AND LONDON

Library of Congress Cataloging-in-Publication Data

Eastern Pennsylvania Branch of the American Society for Microbiology
  Symposium on Infection Control: Dilemmas and Practical Solutions
  (1988 : Philadelphia, Pa.)
    Infection control : dilemmas and practical solutions / edited by
  Kenneth R. Cundy ... [et al.].
        p.   cm.
    "Proceedings of the Eastern Pennsylvania Branch of the American
  Society for Microbiology Symposium on Infection Control: Dilemmas
  and Practical Solutions, held November 3-4, 1988, in Philadelphia,
  Pennsylvania"--T.p. verso.
    Includes bibliographical references.

  ISBN-13: 978-1-4684-5726-1   e-ISBN-13: 978-1-4684-5724-7
  DOI: 10.1007/978-1-4684-5724-7

    1. Health facilities--Sanitation--Congresses.   2. Nosocomial
  infections--Prevention--Congresses.   3. Asepsis and antisepsis-
  -Congresses.   I. Cundy, Kenneth R.   II. American Society for
  Microbiology. Eastern Pennsylvania Branch.   III. Title.
    [DNLM: 1. Infection--prevention & control--congresses.   WC 195
  E131 1988]
  RA969.E2 1988
  614.4'4--dc20
  DNLM/DLC
  for Library of Congress   89-26613
                                                           CIP

Proceedings of the Eastern Pennsylvania Branch of the
American Society for Microbiology Symposium on
Infection Control: Dilemmas and Practical Solutions,
held November 3-4, 1988, in Philadelphia, Pennsylvania

© 1990 Plenum Press, New York
A Division of Plenum Publishing Corporation
233 Spring Street, New York, N.Y. 10013

Softcover reprint of the hardcover 1st edition 1990

ORGANIZING COMMITTEE

CHAIRMAN

Kenneth R. Cundy, Ph.D.
Temple University School of Medicine

CO—CHAIRMEN

Eileen T. Hinks, Ph.D.
United Hospitals, Inc.
Bruce Kleger, Dr.P.H.
Pennsylvania Department of Health

COMMITTEE MEMBERS

Carl Abramson, Ph.D.
Pennsylvania College of Podiatric Medicine
Kathleen Arias, M.S.
Frankford Hospital
Josephine Bartola, J.D.
Pennsylvania Department of Health
Nick Burdash, Ph.D.
Philadelphia College of Osteopathic Medicine
Paul H. Edelstein, M.D.
Hospital of the University of Pennsylvania
Anna Feldman-Rosen, M.S.
Rolling Hill Hospital
Olarae Giger, Ph.D.
Episcopal Hospital
Gary J. Haller, Ph.D.
SmithKline Bioscience Labs
Donald Jungkind, Ph.D.
Thomas Jefferson University Hospital
Linda A. Miller, Ph.D.
Holy Redeemer Hospital & Medical Center
Donald D. Stieritz, Ph.D.
Hahnemann University
George H. Talbot, M.D.
Hospital of the University of Pennsylvania
Norman Willett, Ph.D.
Temple University Schools of Medicine & Dentistry

EX-OFFICIO MEMBER

James A. Poupard, Ph.D.
Medical College of Pennsylvania

PREFACE

When we were setting the theme of "infection control dilemmas and practical solutions" for this symposium, we asked ourselves a basic question: What are some of the most vexing problems and situations facing the hospital microbiologist-epidemiologist team in today's world of opportunistic and new infectious diseases unheard of as common pathogenic occurrences 10 years ago? One of the areas which we immediately focused upon was the tremendous amount of time, energy, and financial resources that are presently being expended to satisfy the requirements mandated by the recognition of the danger of spread of blood-borne pathogens in the hospital environment. With the advent of Universal Precautions, primarily in response to HIV infection and the AIDS crisis, but certainly augmented by the increased incidence of hepatitis in its various forms, a significant effort has been required to meet the standards recommended and/or required by OSHA and the CDC.

With this in mind we brought together experts in the field of infectious diseases to address the problems engendered by the threat of nosocomial spread of selected pathogens. Further, we devoted several sessions to discussing the investigation and resolution of institutional outbreaks of disease, particularly with reference to methicillin-resistant Staphylococcus aureus (MRSA). Special problems of dental offices and clinical teaching as well as extended care facilities were also selected for attention, particularly with relation to blood-borne pathogens.

In light of the tremendous impact that cost containment and reimbursement has had upon the hospital budgetary process, we also attempted to have experts in the field consider in greater detail the various guidelines governing compliance with Universal Precautions. Choices of antiseptics and disinfectants, satisfaction of educational and monitoring activities, and finally the real and perceived risks to health care workers as well as preventive prophylaxis remained uppermost on our agenda when looking at Universal Precautions.

The infectious waste collection, storage and disposal "train" was not the least of the concerns addressed by this symposium. This overlay to the considerable problems already emphasized in previous presentations laid the groundwork for a final presentation of the legal aspects of infection control. This complex legal subject involving all levels of hospital management, direct patient care and ancillary services has great impact upon how workers and institutions respond to the rules, regulations and ethical issues governing delivery of health care.

We do not claim to have solved all of the problems raised in this symposium; and some remain dilemmas. However, we hope to have heightened awareness and provided the reader with pathways for addressing and solving some of these issues as we look to yet another decade of emerging diseases, their recognition and control.

Kenneth R. Cundy

ACKNOWLEDGEMENTS

The editors are grateful to the Eastern Pennsylvania Branch of the American Society for Microbiology for sponsoring this symposium and for making this publication possible. We especially thank the Symposium Committee for their diligent work in organizing an informative and successful symposium.

We would like to acknowledge the support and sponsorship of the Bureau of Laboratories of the Pennsylvania Department of Health, Temple University School of Medicine, Hahnemann University, The Medical College of Pennsylvania, the University of Pennsylvania School of Medicine, Thomas Jefferson University and the Philadelphia College of Osteopathic Medicine.

This symposium would not have been possible without the financial support of the following companies: Beecham Laboratories, Med X Services of PA, Merck, Sharp and Dohme, Roche Diagnostic Systems, Fisher Scientific-IEC, and Becton Dickinson, and Burroughs-Wellcome. We are grateful for their contributions.

Special thanks are offered to Josephine Bartola and the Pennsylvania Department of Health-Bureau of Laboratories for generously providing expertise and assistance with mailing and registrations; to Anna Feldman-Rosen and Lori Walsh for helping in the collection of manuscripts; and to Greg Harvey and Donna May for their assistance in the preparation of some of the camera ready copies for this publication.

Kenneth R. Cundy
Bruce Kleger
Eileen Hinks
Linda A. Miller

CONTENTS

PEDIATRIC INFECTION CONTROL -- AN HISTORICAL PERSPECTIVE

Donald A. Goldmann

Infection Control Program, Department of Medicine
The Children's Hospital, Harvard Medical School
Boston, MA

The recent publication of a history of the Children's Hospital, Boston (1), coupled with the approach of my fifteenth anniversary as hospital epidemiologist at that venerable institution, has prompted me to search for lessons in the Hospital's past, to critically examine the present situation, and to set priorities for the future.

Today, everyone understands that hospitalization of a child has its risks, not the least of which is nosocomial infection. But this was not always the case. When the novel idea of opening an institution exclusively devoted to the care of Boston's children was first broached in the middle of the nineteenth century, the hospital was not viewed by its founders as a potential source of contagion. On the contrary, it was seen as a refuge from the pestilence in the surrounding urban area. As noted by a special commission appointed by the Boston Board of Health to "investigate the sanitary conditions of the city:" (2)

> "Where the population is most dense...and where there are the most foreigners, the massacre of the innocents takes place on the largest scale."

The commission observed that the greatest carnage occurred during the summer months and went on to ask,

> "What can be the nexus between urban density of population and extreme summer heat? It may be inferred that excessive temperatures...among the poor in cities act indirectly... through some intermediate agency. This agency, peculiar to crowded cities...and developing destructive action under the ripening influence of...heat is <u>filth</u>. Uncleanliness consists in the non-removal of...refuse of all kinds... Here, the water...is pure, being brought from a great distance...but the air is impure, especially...in densely populated district...where the sun rarely penetrates,

and ventilation is imperfect...[There is] inadequate provision for removal of the...excreta of a superabundant and habitually unclean population...The air is permanently laden with effluvia from skin and lungs..., noxious gases and vapors from sinks, nooks, crannies, cesspools, and... floating molecules from choked up privies, drains, and sewers. Then..., the ripening action of our midsummer heat kindles, as it were [such air] into a blaze of poisonous putridity."

Ventilation and fresh air were considered paramount in both the prevention and treatment of infectious diseases. Thus, Francis H. Brown in a treatise to the Medical Faculty of Harvard College in 1861 stated that (3):

"Abundance of air; abundance of sunlight; [and] simplicity of construction are the essentials without which no hospital can exist."

As a founder of the Children's Hospital eight years later, Brown was able to put these principles into effect, especially in the addition of "sun balconies" to the hospital in 1901 and in the innovative open pavillion plan for the new hospital upon its relocation to its present Longwood Avenue site in 1914 (Figure 1). The large central courtyard provided space for the annual visit of the circus (Figure 2), as well as a small building for airing out germ-laden mattresses. Children suspected of having tuberculosis were sent to the "country" (Wellesley Hills) where they were housed in 20 X 40 foot "shacks" with doors that were left open all day and often into the night to provide their occupants with the full benefits of fresh air (Figure 3).

The founders of Boston's Floating Hospital took the argument for fresh air one step further. To escape the fetid air and ravages of dysentery in Boston during the summer months, a clinic and small in-patient facility were set up on a boat (Figure 4), which was modelled after a maritime hospital in New York City. From 1894 until it burned in 1927, this hospital ship plied the waters of Boston Harbor with its precious cargo of sick children.

Of course, no matter how airy the hospital environment, the frequency of cross-infection could not be ignored for long. When Dr. Kenneth Blackfan became Chief of Pediatrics in 1923, house officers and students arriving at The Children's Hospital were presented with a booklet describing the management of communicable diseases in patients who had been admitted to hospital for other conditions. If measles, mumps, varicella, or pertussis were suspected, the ward was to be closed to all who were susceptible by history. Instructions were also provided for handling patients with scarlet fever, diptheria, and smallpox. For each of these infections, the house officer was to "order individual isolation and notify Dr. McKhann or [the] Medical Resident (4)."

Before long, confidence in the salutory properties of hospital air was replaced by fear of airborne infection. Abandoning the discredited notion that miasms were responsible for the spread of infection, Wells demonstrated in the 1930s that pathogenic organisms could be aerosolized on droplet

Figure 1. The new Children's Hospital, 1914. Five
Pavillions were connected by walkways.

Figure 2. The circus.

Figure 3.   Children with tuberculosis.

Figure 4.   The Floating Hospital, 1924.

nuclei (5), which are the tiny residual particles left behind
when small droplets evaporate. Bacteria could also be dispersed
into the air on skin squames shed by patients and personnel
or by aerosolization of contaminated dust. Wells' work was
remarkably influential and led to extensive efforts to control
airborne transmission of infectious diseases, not only in
hospitals, but also in public buildings and schools. Antiseptics were sprayed in the air with abandon ("fogging"),
while Wells and his pupils extolled the virtues of ultraviolet
light, which they had demonstrated could kill organisms suspended on airborne droplet nuclei. Not to be outclassed in
the race to adopt the latest medical advances, The Children's
Hospital installed ultraviolet lights in the corridors of
its new infectious diseases ward in an attempt to stop the
spread of varicella, measles, and other airborne infections
(Figure 5) (6). Ultraviolet lights were also placed in operating suites in a number of medical centers, including some
Harvard hospitals.

    Although the efficacy of air decontamination proved
difficult to establish, and fogging quickly fell into disrepute,
ultraviolet light still has it advocates, including some physicians at The Children's Hospital. During my first year as
hospital epidemiologist, I was asked to comment on a proposal
by the Chief of Pediatrics to consider placing ultraviolet
lights in patient care areas. I found myself debating Dr.
Richard Riley, a disciple of Wells, who insisted that cross-
infection on infant wards could be reduced dramatically by
strategically placed ultraviolet lights. Soon thereafter,
the first of numerous annual appeals for ultraviolet lights
in the neurosurgical operating room reached my desk. The
proponents of ultraviolet light seem to have abandoned their
quest for now, but the search for microbe-free air has merely
taken on a more high-tech look as laminar airflow rooms and
"greenhouse" surgical suits have gained in popularity.

    Although the dangers of contaminated air received and
greatest attention four or five decades ago, the role of
hospital personnel and visitors in the spread of infection was
not ignored. The open feeling and free access that characterized the early days of the hospital were replaced by ritual
and regulation. The precautions taken to guard against the
spread of infection by direct and droplet contact were nearly
as extreme as the measures designed to interrupt airborne
spread. Gowns were de rigueur for doctors and nurses, and
masks were often worn as well (Figure 6). Visitation was
strictly controlled, especially in the newborn nursery where
babies entered a faceless antiseptic world virtually devoid
of contact with family members. These rules were enforced
by a powerful hierarchy of stern nurses who brooked no insub-
ordination by any physician, whether chief-of-staff or intern.
No one dared scoff when the head nurse insisted that if a
chart fell on the floor, it had to be picked up with a glove
and aired out (Dr. Willian Berenberg, personal communication).

    It is easy to deride some of these rather naive and anti-
quated views in retrospect, but hospital administrators of
that era had a surprisingly comtemporary outlook on the impact
of nosocomial infections. For example, the opinions of the
Medical Superintendant of the North-Eastern Hospital in London
expressed in 1935 have a familiar ring to today's hospital

Figure 5. Corridor with ultraviolet lights,
           isolation ward, The Children's
           Hospital, 1931 (6).

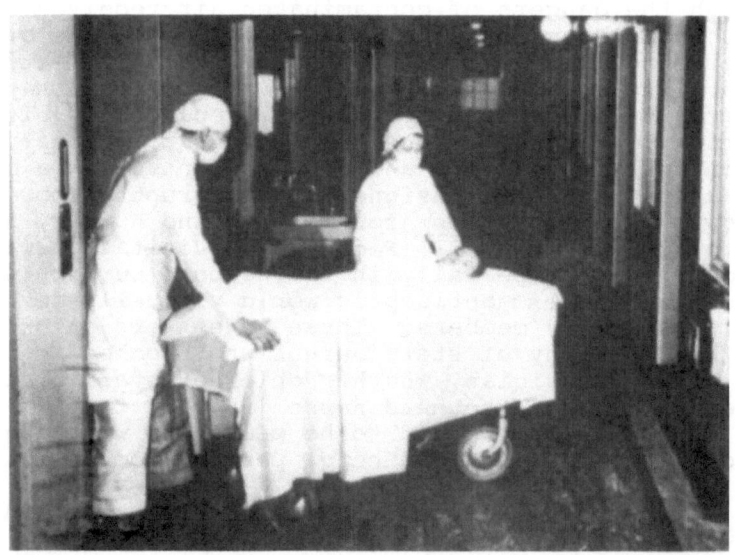

Figure 6. Suitably clothed hospital personnel,
           The Children's Hospital.

epidemiologist obsessed with quality assurance and diagnostic related groups. He noted that hospital acquired infections

> "may be perilous to the individual child; they are always, in some degree, detrimental to the hospital. Long detention, especially if the cause be some superimposed infective process, may result in that complex of psychical deterioration known as hospitalism. From the administrative standpoint, if wards are encumbered by long-stay patients or closed to new ones, the machinery of admission and discharged remains undiminished, [and] the average cost of the patients treated rises. Equally important, the reputation of the individual institution...inevitably suffers."

The staphylococcal pandemic of the 1950s and 1960s had an enormous impact on hospitals, particularly on surgical services and in the nursery. If anything, this worldwide epidemic, with its attendant morbidity and mortality, sharpened the debate between those who viewed airborne microorganisms as the principle threat to hospitalized patients and those who saw good handwashing and barrier precautions as the path to preventing the vast majority of nosocomial infections. As evidence accumulated favoring the latter point of view, modern hospital infection control programs began to take shape, first in Great Britain and then in the United States. In this country, the late 1960s and early 1970s found the Centers for Disease Control providing the intellectual foundation and practical tools for hospital infection control, the Joint Commission on Accreditation of Hospitals codifying infection control requirements, and the Association for Practitioners in Infection Control nurturing the footsoldiers of a new profession.

Such was the situation when I finished my training in the Hospital Infections Branch of the Centers for Disease Control, completed my internal medicine residency program and infectious diseases fellowship, and embarked on my career as a hospital epidemiologist. At first, I struggled to accommodate to the idiosyncrasies of pediatric infection control with its emphasis on nursery infections, viral exanthems, and upper respiratory tract infections. One thing was clear, however: the atmosphere at the Children's Hospital had changed radically in less than a decade. The institution as a whole seemed to have adopted a much more relaxed attitude about the potential dangers of nosocomial contagion. In a sense, the driving philosophy behind the hospital's policies and procedures had more in common with the late 19th century than the middle of the 20th century. The childhood infections which had caused so much concern were either conquered or in decline. Polio, measles, rubella, and diptheria were largely controlled by vaccines, the staphylococcal pandemic had waned, and streptococci and their dreaded sequelae, rheumatic and scarlet fever, had mysteriously lost their punch. Tuberculosis and pertussis, while still a cause for concern, generated only occasional crises, not constant fear.

So the wards were opened up again. Infectious disease units were closed and single bed rooms were abandoned in favor of multiple bed rooms so that hospitalized children could experience a sense of community and companionship. Playrooms

were established on every ward, and the physician who thought
it best to keep a child out of the playroom to reduce the
risk of cross-infection had to run a gauntlet of aggressive,
somewhat skeptical "activities therapists." Hands-on treatment
was emphasized, and it was argued that newborns would have
normal psycho-social development only if touched by human
skin, not gloves. Visitors were encouraged, and brothers
and sisters flooded onto the wards virtually without screening
because it was believed that visitation was critical to the
mental health of both patients and sibblings.

During this period the responsibilities of the infection
control team leaned heavily towards surveillance and epidemic
prevention, detection, and control. There were the usual
pediatric disasters -- unrecognized pertussis on the infant
ward, children whose exposure to chickenpox was not recognized
until <u>after</u> their elective surgery, frequent exposures of
emergency room personnel to patients with meningococcal disease,
an occasional visiting grandmother with cavitary tuberculosis,
and hordes of young visitors on the wards with hacking coughs,
runny noses, runny stools, or funny rashes, scabies in the
phychiatric unit, influenza on the cystic fibrosis ward, and
gastroenteritis in already malnourished babies. Gram negative
nosocomial infections were a persistent problem in the neonatal
intensive care unit. The cardiac surgical service called
in an outside consultant who announced that an increase in
staphylococcal post-operative infections could be attributed
to nurses wearing mini-skirted scrub suits and shedding
staphylococci from the perineum. An aminoglycoside-resistant
strain of <u>Klebsiella pneumoniae</u> -- dubbed "Killer Klebs" by
the surgeons -- was introduced by a patient transferred from
Puerto Rico and meandered about the infant surgical ward for
months. The orthopedic service experienced a small cluster
of Group C streptococcal infections which were traced to an
anal carrier on the surgical staff (7). Shortly thereafter --
and to the amazement of the orthopedic surgeons -- an anal
carrier of Group A streptococci was incriminated in a cluster
of post-operative infections and scarlet fever (8).

Our greatest challenge during this period turned out
to be a most plebeian scourge of the pediatric wards, varicella.
An outbreak of chickenpox on our infant and toddler ward
demonstrated the importance of post-discharge surveillance
and the need to have an open mind when thinking about how
pathogens are transmitted in the hospital (9). For those
of us who tended to discount the role of droplet nuclei in
spreading hospital infections, this outbreak was a rude awak-
ening and a taste of greater problems to come.

A three year-old girl who had been placed on high-dose
corticosteroids for transverse myelitis developed severe
chickenpox and varicella pneumonia while in the hospital.
Fully covered and wearing a mask, she was immediately
transferred to the designated isolation room on the infant
and toddler medical ward and was placed on a ventilator. She
died eight days later. Although complete precautions had
been observed faithfully during her stay, secondary cases
began to occur on the ward about ten days later. Most of the
cases were discovered by our alert infection control practitioner who called the families of all patients who had been

recently discharged. In all, 13 of 24 children who were not immune to chickenpox by serological testing and were on the ward at the same time as the index case became ill. In addition, the mother of another child and a respiratory therapist who never had seen the index patient developed chickenpox. Pediatric cases occurred in all but two rooms on the floor, with some rooms having extremely high attack rates (Figure 7) (9). The child in room C was of special interest. She was nearing discharge after months of total parenteral nutrition for short bowel syndrome when she developed nosocomial <u>Salmonella</u> gastroenteritis, presumably from a child with community-acquired salmonellosis across the hall. She remained on gown and glove enteric precautions for the entire time that the index patient was on the ward, yet she developed chickenpox.

Although transmission of varicella is generally believed to require relatively prolonged, close contact, the epidemiological evidence strongly suggested airborne spread. We hypothesized that the index patient's varicella pneumonia, coupled with her assisted ventilation, had contributed to unusually intense aerosolization of the virus, and we set out to demonstrate how airborne spread might have occurred on the ward. To our surpise, we discovered that our isolation room was at strong positive, rather than negative, pressure with respect to the corridor. This finding sent our engineers scurrying to find the plans for the air handling system of the 20 year-old ward and led to the discovery that an air exhaust in the index patient's room was inoperative. With the door closed, air rushed out under the crack beneath the door with force enough to extinguish a match.

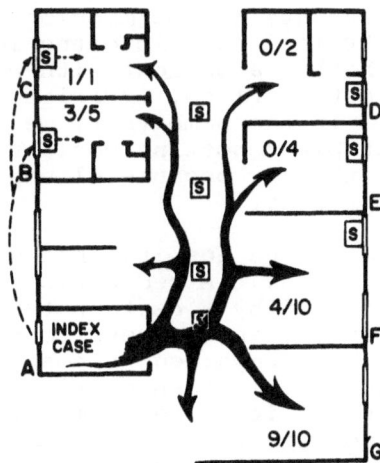

Figure 7. Room-specific varicella attack rates (no. cases/no. susceptibles) and flow of tracer gas (9)

We then used a low-tech approach to trace the airflow
on the ward. We opened a bottle of oil of wintergreen in the
room and soon discovered personnel at other locations on the
ward reflexly rubbing their muscles. Using a more sophisticated
technique, we then released sulfur hexafluoride and followed
its dispersal using infrared spectrometry. The tracer gas
appeared promptly and in high concentrations in room G, where
most of the secondary cases had occurred, but late and in
low concentrations down the hall in rooms B and C (Figure 7).
However, when the index patient's room door was sealed with
tape and the windows were opened, as they had been during
her hospital stay, tracer gas apparently was funelled down
the outside of the building and was sucked into rooms B and
C via through-the-wall heating, ventilation, and air conditioning (HVAC) units. A similar phenomenon was observed in
a hospital in Meschede, Germany in 1970, when smallpox developed
in patients on the floor above the room of the index case (10).
Wells and Riley would have loved it.

Just as I was beginning to acclimatize myself to the
pediatric world, I began to sense that fundamental changes
were occurring in pediatric practice in tertiary centers such
as the Children's Hospital -- changes that would have a profound
impact on infection control. The infection control problems
which we confronted day after day increasingly ressembled
problems faced by my hospital epidemiology colleagues in adult
institutions. It was clear that this trend was merely a re-
flection of changes in the hospitalized pediatric population.
The premature newborns in the neonatal intensive care unit
were becoming ever tinier, our pediatric intensive care unit
patients were sicker and more complex, and many of our children
were increasingly immunosuppressed. Invasive diagnostic and
therapeutic techniques were more numerous and adventuresome,
our surgical procedures were more aggressive, and our transplantation program progressed from kidneys to livers, hearts,
and bone marrows. These changes were reflected even in the
seemingly trivial questions that the practitioner carrying
the beeper has to answer by the score every day. Typical
1975 beeps included requests for permission to bring Lassie
onto the wards and a trained seal into the lobby. In contrast,
in 1985 I was asked whether a patient with a bone marrow transplant could have his pet hairless, irradiated mouse in his
laminar airflow room. In the same year, we informed the crest-
fallen volunteers that plant potting could no longer be considered suitable play therapy because clouds of Aspergillus
were produced in the process.

Even the ambiance of the hospital has changed. The kids
are still special, of course. A pediatric hospital will always
have an atmosphere softened by a certain tender innocence.
But there is a new intensity and aggressiveness that a decade
or two ago would have been regarded as positively un-pediatric.
Our residents have sensed this change, although I am not sure
that they fully understand it. They complain that their clinical experience does not fit their preconceived notions of
what pediatrics ought to be. Their schedule is filled with
intensive care rotations -- neonatal intensive care, pediatric
intensive care, bone marrow transplantation, solid organ transplantation, and "clinical research" (i.e. rare immunodeficien-
cies). Where are the children with the good old pediatric
diseases, they ask? The fact of the matter is that they are

being seen in the ambulatory department, driven off the wards by the reimbursement system and improving diagnostic and therapeutic technologies. So we have come full circle, in a sense, back to the 1930s. Although we are no longer pre-occupied by most of the infections that so concerned Wells and his colleagues, we now realize that airborne chickenpox is a threat to our increasingly immunosuppressed patient population. Moreover, there are new dangers in the air, including nosocomial Legionella and filamentous fungi. We have rejected ultraviolet lights and fogging, but we seem to be embracing high efficiency particulate air (HEPA) filtration and laminar airflow rooms.

The fungi have been particularly nettlesome, and more than any other pathogens, they have been responsible for radical changes in our view of the hospital environment and air supply. One of our initial encounters with fungi came in the neonatal intensive care unit. A critically ill 1,300 gram infant with severe acidosis and hyperglycemia was noted to have a small papule on the cheek. This lesion rapidly evolved into an extensive necrotic ulcer with a shaggy base which continued to expand despite debridement. Biopsy showed invasive fungus, and culture grew Rhizopus. The baby died only a few days after the infection was first noticed.

Upon questioning the intensive care unit nursing staff, it was evident that the papule had formed in an area of the skin that had been covered by tape used to secure the baby's endotracheal tube. The roll of tape was retrieved, and culture revealed Rhizopus. However, further investigation demonstrated that the tape was an innocent bystander in an environment that was heavily laden with Rhizopus. Fungus-laden dust was found everywhere in the unit, particularly in the air exhaust duct near the infected patient's radiant warmer. Fluffy dust balls were found around and under monitoring equipment. Air sampling documented heavy air contamination with fungal spores even though the near-HEPA filtration system was functioning normally. Although the ultimate origin of the problem could not be defined, it seemed clear that inadequate cleaning practices, particularly the reluctance of housekeeping personnel to disturb delicate electronic equipment and the limited time available to clean such a busy unit, contributed to the persistence of Rhizopus in the environment. Although renovations were underway in the adjacent ward, Rhizopus was not found in that area.

And why were we renovating the adjacent ward? Fungi again, this time Aspergillus flavus. Since 1983 sporadic cases of A. flavus infection had been occurring on this bone marrow transplantation/oncology ward. At first, there were very few infections, but when three infections were noted in a two month period in 1984, our concern grew. Shortly before this cluster occurred, the old credit union next to the hospital had been imploded. We found numerous routes by which fungal spores liberated in the process could have gained access to the ward. Air was supplied by a very inefficient bag filtration system, which certainly was incapable of screening our fungal spores. Windows were opened frequently, and even when closed leaked prodigious amounts of air. Through-the-wall HVAC units provided a clear passage for fungus-laden air. In addition, the inside of the HVAC units, especially the fans, was heavily contaminated with dust. Cultures revealed a heavy growth of A. flavus and other pathogenic fungi. A hung ceiling provided an additional reservoir for spores.

As a result of these alarming findings, the ward was closed and underwent renovations costing in excess of $300,000. The ceilings were plastered, the HVAC units were removed, the windows were sealed, and a HEPA filtration system was installed. The expense and disruption to patient care were particularly difficult for the hospital administration to swallow because digging had just begun for a new patient care facility. Nevertheless, an additional investment of nearly $200,000 was made shortly thereafter to consolidate the bone marrow transplantation beds in laminar airflow rooms. These rooms were purchased not out of any conviction that they were known to improve the outcome of transplantation, but rather to provide a further measure of security in an antiquated structure. The one positive aspect of this crisis was that it provided us with an invaluable dry run in designing wards for high risk patients in the new hospital.

Then to our amazement, two cases of fungal sinus infection (one documented to be caused by A. flavus) occurred in a period of two weeks in the same room. Both patients were teenagers who had undergone induction therapy for acute myelogenous leukemia and were profoundly neutropenic. These infections were particularly alarming, not only because of their close temporal association, but also because they occurred so soon after induction and presented with strikingly similar clinical pictures. The inescapable conclusion was that there had to be an environmental source. An alert infection control practitioner directed our attention to the ward's elevator shaft. The shaft was clogged with Aspergillus-laden dust. As the elevator ascended the shaft to the top floor, which housed the transplantation/oncology unit, it acted as a piston, driving the dust before it. Elevator cables clanged and scraped against the walls, liberating still more dust. A torn sheet of plastic, which was designed to keep dust out of the machinery in the penthouse above, flapped about, aerosolizing clouds of fungus. Air sampling revealed Aspergillus in the air near the elevators, which were, of course, used by patients on the way to diagnostic procedures. In addition, dust was detected trapped in the grids covering the new light fixtures throughout the unit, as well as around some of the air vents, apparently having been attracted by airflow and, perhaps, electrostatic forces. The seemingly harmless innovative light diffusers turned out to be fungus collectors dangling their deadly burden directly above the beds of immunosuppressed patients. They were difficult to clean, but the housekeeper had found a way. He had rolled up a Red Sox pennant and used it to clean out the dust, hole by hole.

Of course, no review would be complete without acknowledging the impact of AIDS on the pediatric hospital. AIDS has brought a new level of immunosuppression and complexity of care to our wards. However, the impact on our staff may reveal more about from whence we have come and where we are going as health care providers. Not long ago, a would-be doctor or nurse faced a substantial risk of contracting tuberculosis, polio, or other crippling and potentially fatal diseases. A positive PPD came with the turf. In more recent years, however, the hazards of clinical duty have become much less obvious and considerably less serious. Few nursing or medical students would have characterized their chosen trades as dangerous. Thus, when the AIDS epidemic unfolded, the Public Health Service seemed intent on reassuring the medical community that the risk to health care providers was virtually non-existant. When cases did occur, no effort was spared to prove that the affected individuals belonged to other high risk groups and probably did not acquire their infection on-the-job. Of course, hospital acquired cases in health care workers had to occur eventually, and when they did, there was a mad scramble to put them in context and to reassure physicians and nurses yet again.

The result was universal precautions. This is not the place to discuss the pros and cons of this controversial policy. However, universal precautions appears to be having a para-doxical effect on the core of modern infection control practice -- our belief that the best way to prevent the spread of infection is to wash hands or practice barrier precautions when moving from patient to patient. In response to universal precautions, personnel are wearing gloves, particularly when they have reason to believe that a patient may be in a high risk group. But the gloves are being worn to protect the caregiver, not other patients. My colleagues and I have found that it is not uncommon for doctors and nurses to wear the same pair of gloves while caring for multiple patients, oblivious to the fact that while protecting their own hands, they are transporting traditional nosocomial pathogens around the ward (11).

The changes I have observed at the Children's Hospital centainly are not unique. Similar trends are evident through-out the nation. How should hospital epidemiologists respond to these daunting challenges and complex problems? Within the professional organizations, the Society of Hospital Epidemiologists of America and the Association for Practitioners in Infection Control, there are calls for a broader definition of our mission. It has been suggested that we exploit our skills and training and insure our future livelihood by cap-turing the hot new market in quality assurance and risk management. There can be little doubt that these fledgling disciplines would benefit from the skills and experience of hospital epidemiologists, and those who have an interest in these areas certainly should answer the call. But I would suggest that infection control offers challenges enough, and that for many, quality assurance and risk management will be distractions, robbing us of the energy and time we need to innovate and perform our jobs with distinction. Rather than expanding our role, I would address the following priorities.

First, hospital epidemiologists should be true to their name or call themselves something else. The vast majority of so-called hospital epidemiologists have had no formal training in their discipline (12). To be effective, the epidemiologist does not need to be a crack biostatistician, but he or she does need to know how to collect valid data, to use relatively straightforward techniques to assess the risk factors for and the consequences of nosocomial infection, and, based on these analyses, to design and evaluate practical strategies. We are well past the stage when we can justify our existence merely by counting infections and instituting control measures recommended by some Federal agency.

The problem, of course, is to define a relevant problem that can be attacked with existing resources within the walls of a single institution. After all, collaborative studies are difficult to organize, and funding historically has been hard to obtain for infection control projects that do not involve an antibiotic or commercial device. At the Children's Hospital, we have chosen to place a major research emphasis on one of our most important clinical problems, respiratory syncytial virus (RSV) infection. RSV is a major cause of morbidity and mortality in infants worldwide. Thanks to the pioneering work of Dr. Caroline Hall and her colleagues in Rochester, we now know that RSV is a significant problem in hospitalized infants as well. Indeed, Hall's work could serve as an inspirational model for us all, since she has described the nosocomial RSV problem, quantitated its magnitude, defined its epidemiology, and proposed and evaluated several potential control measures. Hall's studies would not have been possible without the revolution that has taken place in diagnostic virology in recent years. Insensitive, labor-intensive procedures that were suitable only for reference centers have been replaced by rapid, commercially available kits that can detect viruses directly in clinical specimens.

Hall demonstrated that seasonal epidemics of RSV in the community inevitably are mirrored by outbreaks of nosocomial RSV in the hospital as infants with bronchiolitis are admitted to pediatric wards. The resulting rates of nosocomial RSV infection are truly astounding, approaching 45% among infants hospitalized for one week or longer and 100% among children who remain in the hospital for more than a month (13). These astronomical rates of infection have particularly grave consequences because they occur in a concentrated population of patients who already have serious underlying diseases. Up to one-third of neonates in intensive care suffer nosocomial RSV infection, which tends to be heralded by non-specific signs such as apnea. In one study, 17% of infected neonates died, half of them unexpectedly (14). Significant complications of RSV infection are also seen in children with congenital heart disease (15) and immunosuppression (16).

RSV is transmitted among hospitalized children by direct exposure to contaminated secretions themselves or to large droplets produced by coughing or sneezing, not by droplet nuclei. This was demonstrated in a clever study by Hall's group (17). Adult volunteers were asked to expose themselves to a baby with acute RSV infection. These volunteers were divided into three groups. "Cuddlers" had intimate exposure to the infected patient over a period of two to four hours. "Touchers" had contact with surfaces that were contaminated with the infant's secretions, such as crib rails, pacifiers, and toys, but did not provide direct care. Since RSV requires inoculation onto the mucous membranes of the nose or eyes to produce infection, volunteers increased their risk of infection by vigorously rubbing their nose and eyes with contaminated hands. "Sitters" sat at least six feet away from the infected baby and read a book for three hours. RSV infection occurred in 5/7 "cuddlers," 4/10 touchers," and none of the 14 "sitters," confirming that infection required contact with the infected infant or with surfaces contaminated with secretions containing viable virus.

Thus, RSV can be transmitted either by direct transfer of virus from patient to patient on the hands of personnel. Alternatively, hospital personnel can themselves serve as vectors if they are infected by the infants under their care. Since there is no long-term immunity to RSV, attack rates among caregivers may reach 25% to 50% during the peak of an outbreak (13). Failure to base infection control strategies on these well established mechanisms of transmission can lead to illogical policy. For example, the Centers for Disease Control recommends that infants with RSV be placed in private rooms despite the fact that RSV is not an airborne pathogen and few hospitals have sufficient single rooms to handle the onslaught of infected children each winter. Gowns are also recommended, but unaccountably, gloves are not. Although neither gowns nor masks would be expected to halt contact spread of RSV, these measures have been evaluated in two small studies, neither of which showed a beneficial effect (18, 19). On the other hand, disposable goggles that covered both eyes and nose, accompanied by diligent handwashing, were effective in reducing the rate of infection among patients and staff, at least over a short period of time (20).

We decided to attack the central problem, contamination of the hands of hospital personnel. Handwashing alone almost certainly would be effective, but this simple measure frequently is neglected by busy doctors and nurses. Therefore, we evaluated the efficacy of gown and glove contact precautions, which are routinely used for other nosocomial pathogens known to be transmitted via the hands, such as Staphylococcus aureus. The study was performed on our 28 bed infant and toddler ward (21) and included the three RSV seasons from 1982 to 1985. We arbitrarily defined the RSV season as the 24 week period each year beginning in November and ending in April, so a total of 72 weeks were included in the three year study. Although attempts were made to place infected patients in single rooms or to cohort them in multi-bed rooms, this was rarely possible. Continuous intensive surveillance was performed with the aid of a rapid immunofluorescence RSV detection technique used routinely by the diagnostic virology laboratory. Nosocomial RSV infection was defined as infection presenting at least five days after admission, based on the average incubation period of this disease. Post-discharge surveillance was not performed, but patients readmitted for nosocomial infection were counted. Patients were assumed to be susceptible to RSV unless they entered with community-acquired disease or until they developed a nosocomial infection.

During the course of this study, we measured the effect of an intervention designed to improve compliance with gown and glove precautions. At the start of the study, surreptitious monitoring revealed that nurses were poorly compliant with these precautions and used both gloves and gown for only 38.5% of contacts with infected patients. In the middle of the study period, compliance was openly monitored with the full participation of the head nurse, and this produced the desired Hawthorne effect. Nursing staff wore both gloves and gowns for 81% of their contacts with patients on precautions at the start of the monitoring period, and compliance improved to nearly 100% within a few weeks. To our surprise, compliance remained excellent for the rest of the study, long after open monitoring was suspended, presumably because personnel noticed a decline in the nosocomial infection rate and were constantly motivated by a strong continuing education program that had the full support of the head nurse.

Improved compliance with contact precautions was associated with a dramatic decline in the nosocomial RSV infection rate. After adjusting for the level of exposure to infants on the ward who were excreting RSV during each week of the study, the summary relative risk of nosocomial RSV was 2.9 (95% confidence interval 1.5-5.7), comparing the incidence density of infection in the period before the intervention to the incidence density in the period when compliance with precautions was excellent. In other words, the risk of infection was nearly three times as great when nurses neglected to use gloves and gown. As expected, linear trend analysis revealed a strong overall association between the infection rate and the level of exposure to RSV on the ward. Interestingly, the slope of this relationship was steeper during the low compliance period; the risk of nosocomial RSV infection increased more than four times as rapidly with the level of exposure in the low compliance period than in the high compliance period. That is, the protective effect of precautions was greatest when the number of patients on the ward excreting RSV reached its peak and the challenge to the infection control program was greatest.

The second important challenge to the hospital epidemiologist in the coming decade is to fully exploit recent advances in molecular microbiology, including agarose gel electrophoresis and restriction endonuclease analysis of bacterial plasmids, restriction endonuclease analysis of viral and bacterial chromosomal DNA, and selective probing of variable regions of bacterial plasmid and chromosomal genes (22). Such techniques have been particularly useful in tracing the spread of plasmids and transposons containing antibiotic resistance genes in the hospital environment. Nowhere has molecular epidemiology been used to greater benefit than at the Seattle Veterans Administration Hospital (23,24). In 1976 that institution experienced an outbreak of Serratia marcescens resistant to multiple antibiotics, including gentamicin and tobramycin. Within a few years, aminoglycoside resistance began to appear in other gram-negative bacilli, including Citrobacter, Proteus, Providencia, and Enterobacter, on a number of hospital wards. All of these strains, regardless of genus and species, had a similar zone of inhibition on disc diffusion testing and elaborated a 2" adenylating aminoglycoside inactivating enzyme, suggesting that they harbored the same plasmid. In fact, a common 45 megadalton plasmid was demonstrated by agarose gel electrophoresis -- the infamous pLST 1000 plasmid that subsequently has been found in hospitals in several parts of the world (25). The identity of plasmids in a number of strains was confirmed by restriction endonuclease analysis of plasmid DNA.

Soon, however, the microbiologists and epidemiologists at the Seattle Veterans Administration Hospital were faced with an even more difficult problem. A similar resistance pattern had appeared in nosocomial isolates that carried a new larger plasmid. Therefore, the progress of the outbreak could no longer be followed by the relatively simple strategy of screening for the 45 megadalton plasmid. Spread of anti-biotic resistance genes could not even be detected by assaying for the 2" adenylating enzyme, since the associated plasmid genes had become cryptic in some strains. To facilitate detection of these genes, a probe was constructed which was highly sensitive and specific for the genes controlling the aminoglycoside adenylase, thus providing the investigators with a truly superb epidemiological marker. Similar probes, albeit radiolabelled and not suitable for routine use by the hospital epidemiologist, are being developed for other plasmid-associated antibiotic resistance genes.

The third priority for the hospital epidemiologist is to forge a partnership with investigators who are primarily involved in studying the pathogenesis and treatment of infectious diseases. There are, after all, limits to what epidemiologists can hope to achieve in preventing the spread of infection by modifying behavior of the hospital staff. Since patients invariably are infected by organisms that they have brought with them from the community or have acquired during their hospital stay, the trick is to figure out how these pathogens colonize the patient, overcome his host defenses, and ultimately produce clinical disease. If the colonization process is understood, perhaps it can be blocked. If the virulence properties of a pathogen are elucidated, perhaps they can be neutralized. If the defects in host defense that permit microbial invasion can be defined, perhaps they can be corrected. As basic mechanisms of pathogenesis are clarified and appropriate interventions are developed and introduced into practice, clinical trials will be required. Instead of smugly criticizing the experimental design and analysis of these studies, hospital epidemiologists would be well advised to become involved in the process themselves. Certainly, we are in a position to influence the research priorities of our infectious diseases colleagues and to help them test their theories and discoveries on our hospital wards. Effective collaborations can be established with relative ease, as demonstrated by studies of the pathogenesis of gram negative bacillary colonization of the alimentary tract of neonates and coagulase-negative staphylococcal colonization of prosthetic materials at the Children's Hospital have demonstrated (26,27).

The challenge of controlling hospital-acquired infections has always been daunting. The problems faced by hospital epidemiologists today are even more complex and challenging than they used to be. Fortunately, we have an array of tools at our disposal that are far more sophisticated than those available to our predecessors. All we need to do is use them wisely.

REFERENCES

1. C. A. Smith, Built better than they knew, in: "The Children's Hospital of Boston," Little, Brown and Co., Boston/Toronto (1983).
2. Ibid, p. 10.

3. <u>Ibid</u>, p. 10-11.
4. <u>Ibid</u>, p. 147-148.
5. W. F. Wells, "Airborne Contagion and Air Hygiene," Harvard University Press, Cambridge (1955).
6. C. F. McKhann, A. Steeger, and A. P. Long, Hospital infections. I. A survey of the problem, <u>Amer J Dis Child</u>. 55:579-599 (1938).
7. D. A. Goldmann and S. J. Breton, Group C streptococcal surgical wound infections transmitted by an anal-rectal and nasal carrier, <u>Pediatrics</u>. 61:235-237 (1978).
8. D. D. Richman, S. J. Breton, and D. A. Goldmann, Scarlet fever and group A streptococcal surgical wound infection traced to an anal carrier, <u>J Pediatr</u>. 90:387-390 (1977).
9. J. M. Leclair, J. A. Zaia, M. J. Levin, R. G. Congdon, and D. A. Goldmann, Airborne transmission of chickenpox in a hospital, <u>N Engl J Med</u>. 302:450-453 (1980).
10. Anonymous, Smallpox outbreak, <u>WHO Weekly Epidemiol Rec</u>. 45:249 (1970).
11. B. N. Doebbeling, M. A. Pfaller, A. K. Houston, and R. P. Wenzel, Removal of nosocomial pathogens from the contaminated glove: implications for glove reuse and handwashing, <u>Ann Intern Med</u>. 109:394-398 (1988).
12. D. A. Goldmann, Nosocomial infection control in the United States of America, <u>J Hosp Infect</u>. 8:116-128 (1986).
13. C. B. Hall, R. G. Douglas, J. M. Geiman, and M. K. Messner, Nosocomial respiratory syncytial virus infections, <u>N Engl J Med</u>. 293:1343-1346 (1975).
14. C. B. Hall, A. E. Kopelman, R. G. Douglas Jr, J. M. Geiman, and M. P. Meagher, Neonatal respiratory syncytial virus infection, <u>N Engl J Med</u>. 300:393-396 (1979).
15. N. E. MacDonald, C. B. Hall, S. C. Suffin, C. Alexson, P. J. Harris, and J. A. Manning, Respiratory syncytial viral infection in infants with congenital heart disease, <u>N Engl J. Med</u>. 307:397-400 (1982).
16. C. B. Hall, K. R. Powell, C. L. Gala, M. E. Menegus, S. E. Suffin, and H. J. Cohen, Respiratory syncytial virus infection in children with compromised immune function, <u>N Engl J Med</u>. 315:77-81 (1986).
17. C. B. Hall and R. G Douglas Jr, Modes of transmission of respiratory syncytial virus, <u>J Pediatr</u>. 99:100-103 (1981).
18. C. B. Hall and R. G. Douglas Jr, Nosocomial respiratory syncytial virus infections: Should gowns and masks be used?, <u>Amer J Dis Child</u>. 135:512-515 (1981).
19. D. Murphy, J. K. Todd, R. K. Chao, I. Orr, and K. McIntosh, The use of gowns and masks to control respiratory illness in pediatric hospital personnel, <u>J Pediatr</u>. 99:746-750 (1981).
20. C. L. Gala, C. B. Hall, K. C. Schnabel, P. H. Pincus, P. Blossom, S. W. Hildreth, R. F. Betts, and R. G. Douglas Jr, The use of eye-nose goggles to control nosocomial respiratory syncytial virus infection, <u>JAMA</u>. 256:2706-2708 (1986).
21. J. M. Leclair, J. Freeman, B. F. Sullivan, C. M. Crowley, and D. A. Goldmann, Prevention of nosocomial respiratory virus infections through compliance with glove and gown isolation precautions, <u>N Engl J Med</u>. 317:329-334 (1987).
22. D. A. Goldmann, New microbiological techniques for

hospital epidemiology, <u>Eur J Clin Microbiol</u>. 6:344-347 (1987).

23. L. S. Tompkins, J. J. Plorde, and S. Falkow, Molecular analyses on R-factors from multiresistant nosocomial isolates, <u>J Infect Dis</u>. 141:625-636 (1980).

24. F. C. Tenover, T. D. Gootz, K. P. Gordon, L. S. Tompkins, S. A. Young, and J. J. Plorde, Development of a DNA prove for the structural gene of the 2"-0-adenyltransferase aminoglycoside-modifying enzyme, <u>J Infect Dis</u>. 50:678-687 (1984).

25. T. F. O'Brien, K. H. Mayer, H. Kishi, M. Syvanen, and J. D. Hopkins, Intercontinental spread of a new antibiotic resistance gene on an epidemic plasmid, <u>Lancet</u>. 2:87-88 (1985).

26. N. G. Guerina, T. W. Kessler, V. J. Guerina, M. R. Neutra, H. W. Clegg, S. Langermann, F. A. Scannapieco, and D. A. Goldmann, The role of pili and capsule in the pathogenesis of neonatal infection with <u>Escherichia coli</u> Kl infection, <u>J Infect Dis</u>. 148:395-405 (1983).

27. M. Tojo, N. Yamashita, D. A. Goldmann, and G. B. Pier, Isolation and characterization of a capsular polysaccharide/adhesin from <u>Staphylococcus epidermidis</u>, <u>J Infect Dis</u>. 157:713-722 (1988).

LABORATORY APPROACH TO AN OUTBREAK OF NOSOCOMIAL INFECTION:

SYSTEMS AND TECHNIQUES FOR INVESTIGATION

John E. McGowan, Jr.

Emory University School of Medicine
and Grady Memorial Hospital
Atlanta, Ga. 30335

INTRODUCTION

Dealing effectively with an outbreak of nosocomial infection requires confirmation of the identity of the outbreak, complete case-finding, comparison of current attack rates with usual baseline, characterization of epidemiologic features, development and implementation of control measures, and continuing surveillance to determine if control measures were successful in ending the episode.

The laboratory has crucial responsibilities in carrying out this process. Among these are rapid communication of microbiologic findings that suggest an outbreak, confirmation of the microbiologic diagnosis, provision of records that permit case-finding and allow comparison with unusual occurrence, processing supplementary cultures of patients, personnel, or the environment, assuring that organisms isolated during the investigation will be available for further study, and conduct of additional microbiologic studies to establish similarity or differences among isolates. Systems for data management have made a big impact on recordkeeping requirements, and have facilitated communication between laboratory and others involved in the investigation.

Development of new systems for characterizing some pathogens has facilitated the assessment of organism similarity. However, some of these systems have impeded, rather than helped, outbreak investigations or have produced "pseudoepidemics" of nosocomial infection. Some typing methods provide data of doubtful value or results that are difficult to interpret. The microbiologist must be aware of the potential benefits and drawbacks of these characterization procedures, so that effective, efficient, and economical methods will be chosen.

Hospital outbreaks of infection are a major focus of concern for infection control programs. The need for hospital infection control programs arose from the Staphylococcal epidemics of the period 1950-1970, and outbreaks still justify infection control efforts today. The laboratory plays a key role in dealing with nosocomial outbreaks, and cooperation between laboratory personnel and epidemiology workers is crucial to effective control. This paper discusses the elements of hospital response to an outbreak, then focuses on the laboratory's role in the control efforts.

WHAT IS A HOSPITAL OUTBREAK?

What defines a hospital outbreak? Many answers have been proposed, but one with general applicability is that proposed by Dixon, in which a hospital outbreak is the occurrence of a greater than expected number of cases (Dixon, 1986). By this definition, the baseline occurrence of infections determines the presence or absence of an outbreak. Classically, sharp and unexpected increases in number of cases makes recognition easy. However, this is not always needed. For example, cases of infection due to enterococci with high level resistance to gentamicin have been extremely rare in most hospitals until recently (Patterson et al., 1988). Even one case of infection due to a gentamicin-resistant enterococcus well might be defined as an outbreak, especially in view of the difficulty involved in treating infections due to these organisms (Patterson et al., 1988). On the other hand, several cases of nosocomial infection due to Escherichia coli might not be considered an outbreak if the isolates all had the susceptibility pattern and biochemical reactions characteristic of most strains found in the hospital and if cases due to this organism were encountered frequently at the institution.

HOW OFTEN DO OUTBREAKS OCCUR?

A study by the Centers for Disease Control reviewed occurrence of nosocomial outbreaks in seven community hospitals throughout the United States (Haley et al., 1985). Eight outbreaks were confirmed in one year, involving 82 patients, or 2% of all nosocomial infections and 0.09% of all hospital discharges. Extrapolated to the U.S. as a whole, this rate of occurrence suggests about one outbreak per year for the average community hospital with 150 beds or more. Half of the recognized outbreaks resolved spontaneously; two required that the hospital institute control measures, and the other two were resolved only after an outside investigation. The authors also detailed some evidence that the organisms associated with outbreaks in larger hospitals may be "uncharacteristic" of epidemics in small community hospitals (Haley et al., 1985).

SYSTEMS AND TECHNIQUES FOR DEALING WITH A SUSPECED OUTBREAK

The fundamental approach to dealing with a hospital outbreak of infection involves several steps and phases (Clegg and Dixon, 1979); the laboratory has a role in virtually all of these activities (Table 1). Seven steps will be considered.

1. The initial step of outbreak recognition is the most important. Infection control personnel may become aware of the

Table 1. Steps in Investigation of a Hospital Outbreak and the Laboratory's Role in Each Step

Investigative Step	Laboratory Participation

1. Recognition of the problem
 	
  a. Case definition
  b. Verification of clinical entity
	1. Laboratory surveillance/ early warning
	  a. Microbiologic confirmation

2. Complete case finding
 	
  a. Reliability of reporting
 	
  b. Completeness of reporting
  c. Obtain additional data
	2. Accurate microbiologic characterization
	  a. Search data base for other cases; review lab methods
	  b. Complete demographics
	  c. Process new cultures as needed

3. Is this unusual?
 	
  a. Calculate attack rate
  b. Compare with baseline rates
  c. Epidemic = higher than usual rate
	3. Provide archival data on occurrence

4. Characterize the outbreak
  a. Patient demography
  b. Location
  c. Time
	4. Is this one strain or many?
	  a. Typing of isolates

5. Form hypotheses about causes
 	
  a. Mode of spread
  b. Reservoirs
  c. Vectors
	5. Environmental cultures as needed

6. Initiate control measures	
7. Follow-up surveillance to make sure control measures effective	7. Laboratory surveillance/ early warning

problem through their contact with, and surveillance conducted on, the clinical services. On occasion, however, the laboratory, a clinic, or even another hospital in the area may provide the first report. When a question arises about whether an outbreak is present, or when one is clearly occurring, the investigators must begin by establishing a definition of a case. Even a rough definition will help the laboratory begin to examine microbiologic aspects, allow initial control measures, and assure that all involved agree on the nature of the problem. This will prevent problems like an investigation

of presumed toxic shock syndrome that turns out to be streptococcal scarlet fever, or vice versa.

2. Once a definition has been made, attempts to identify
all possible cases begin. This effort at complete case-finding
is important because the more cases that are available for
analysis, the better the chance of determining the process
involved. Case-finding has three major aspects. The first of
these is ascertaining the reliability of both clinical and laboratory information. Can the clinical staff identify a case of
the entity under investigation? Has the laboratory experience
in recognizing the microbiologic entity being considered? This
is the stage where "pseudooutbreaks" (see below) need to be
weeded out. Next, the completeness of reporting of cases must
be considered (Bartlett, 1987). For example, a number of
studies have shown that up to half of all surgical wound infections become manifest after the patient has left the hospital
(Weigelt et al., 1988). Can the reporting of cases be improved
by special attention to surveillance measures? When intense
search for all cases is completed, the completeness of the
information obtained about each case must be reviewed. For
example, if most of the information on cases comes from the
laboratory, the demographic information may be lacking, as it
seldom is provided with requests for laboratory specimens. At
Grady Memorial Hospital, our laboratory requisition asks the
ward on which the inpatient is found, but not the clinical
service of the patient. Thus, our laboratory would be unlikely
to recognize a service-specific outbreak.

3. When as many cases as possible are identified, the
question of whether this episode meets the definition of an
epidemic again can be considered. The attack rate of current
cases must be compared with the usual, "baseline" occurrence of
the entity at the institution. Presence of an attack rate
higher than usual then confirms that an outbreak is taking
place.

In these first three steps, the most important role for
the laboratory is accurate identification and susceptibility
testing of suspected outbreak organisms. Most organisms
causing outbreaks of nosocomial infection can be cultured.
Identification of the etiology of an outbreak can depend on the
extent to which the laboratory carries out characterization of
the organism(s) in question, both as a routine and in special
outbreak circumstances. Some hospitals do not have the
resources to speciate certain organisms in routine practice
(Coyle, 1984). In an outbreak, these institutions may have to
begin to speciate relevant isolates, or to retrieve relevant
strains for speciation, as circumstances demand. At a minimum,
the laboratory should be capable of identifying gram-positive
cocci and gram-negative aerobic bacilli to the species level
when special or recurring epidemic problems make this necessary. Of course, retrospective testing of strains is possible
only when isolates of possible epidemiologic importance have
been saved systematically by the laboratory; the resources to
do this vary markedly from hospital to hospital, so it well
might be that prior isolates are not available. Timely reporting is also a crucial responsibility of the laboratory
(Parkhurst et al., 1985). As soon as it is apparent that a
problem exists, all relevant organisms should be reported to
infection control personnel as soon as their isolation is

recognized. Arrangements should be made to save all potentially relevant organisms at least for the duration of the
investigation, and perhaps longer, if storage facilities allow.
Archival information must be maintained by the laboratory to
establish the baseline of occurrence of the diseases in
question. Ironically, archival information required for
epidemic investigation may be harder to retrieve now than
before the advent of electronic data processing systems, as
information on biochemical reactions, etc., traditionally saved
on the laboratory worksheets may not be part of the record
maintained on the modern Laboratory Information System.

4. Characterizing the outbreak in terms of its location,
time frame, persons, procedures instruments, or other features
is the next step in the systematic approach. In the presence
of enough cases, epidemiologic patterns may emerge here that
allow the problem to be understood. Outbreak investigation may
require identification not only of patients but also of personnel who may be colonized with the outbreak strain. Special
techniques may be required to accomplish such projects. For
example, reliable detection of Salmonella carriage may require
enhancement of growth by use of selective media. The laboratory must make careful assessment of the sites to be cultured
and of the culture media and techniques to be employed
(McGowan, 1985). Likewise, the laboratory should be prepared
for processing environmental cultures to help determine or
confirm the mode of spread. While environmental cultures are
not recommended on a routine basis in the hospital setting,
specific culture surveys to determine mode of spread in known
cases of patient illness are a major contribution in many outbreak investigations. For example, in an outbreak of infection
with enterococci demonstrating high-level resistance to gentamicin, personnel and environmental cultures established that
the mode of spread to additional patients was from person to
person on the hands of personnel (Zervos et al., 1987). Barrier
precautions to limit such spread were a major factor in controlling the epidemic. When environmental or personnel cultures
are needed, they should be considered one of the costs of the
hospital's infection control program, and not billed to the
individual patients affected in the outbreak. Culture techniques for a wide variety of items and substances have been
described (McGowan, 1985).

5. Developing hypotheses as to the reasons for the
outbreak, the reservoir of the organisms, and the mode of the
agents spread to the patient are goals of the next step in
the process. Such theories usually point to certain control
measures that can be taken to check the progress of the
epidemic. For example, if the data suggest that the air-
conditioning system of the operating suite is contaminated
with the epidemic organism, the actions needed to minimize
focus on improving air cleanliness rather than on handwashing.
In developing hypotheses, the medical literature is often a
good place to start - other institutions may have had similar
problems with a given organism, and worked out reservoir and
mode of spread. For example, in the outbreak listed above of
infection with gentamicin-resistant enterococci, it was postulated that selection of resistant organisms from heavy use of
broad-spectrum antimicrobial agents and spread of the resistant
organisms from patient to patient on the hands of employees both
led to the outbreak (Zervos et al., 1987). Testing certain

hypotheses may require the laboratory to define whether the outbreak strains are related or unrelated to each other. A number of typing systems to determine organism similarity or difference have been developed; these will be considered at greater length below.

6. Actions to control the outbreak now can be designed that are based on the analysis completed at this point. On the basis of the new understanding, new steps should be implemented, or efforts taken at the outset should be revised. For example, control of high-level gentamicin-resistant enterococci is said to require attention to use of broad-spectrum antibiotics and attention to aseptic techniques of barrier isolation (Hoffman and Moellering, 1987).

7. Follow-up surveillance must be conducted to make sure that the control measures achieved the desired effect. In this step the laboratory retains a prominent role, as a potential early warning system for recurrent cases if control measures are ineffective.

TYPING SYSTEMS FOR OUTBREAK INVESTIGATIONS

On occasion, the epidemiologist will suspect a common origin for the isolates in an outbreak. The laboratory then is asked to determine if organisms are similar or different. Such studies are a key to determining the likely causes of the episode. For example, assume that a number of cases of staphylococcal infection are seen in patients who have a specific type of surgery. All isolates are of the same genus and species. If laboratory studies can show that these isolates represent one strain that has become widely disseminated, this would suggest a common vehicle of infection. By contrast, typing studies showing the presence of many different strains of the same organism would suggest a breakdown in usual aseptic practices. Since control measures for each of these situations vary, the laboratory contribution of strain typing can be extremely important in determining the action that will be taken to deal with an outbreak.

The usual initial attempts to investigate whether the organisms are similar or different examine the results of biochemical tests and pattern of susceptibility to antimicrobials. However, some common nosocomial pathogens will have similar results of these analyses on the basis of chance alone, and other organisms characteristically have such small differences in these characteristics from strain to strain that the tests provide little information about similarity or difference (McGowan, 1985). Also, some organisms will gain or lose plasmid or chromosomal genes, resulting in an apparent shift of pattern. In these situations, examination ("typing") of further organism characteristics ("markers") can provide data of great epidemiologic consequence.

A variety of new typing systems have been developed over the past decade (Farmer, 1988; Goldmann, 1987; Hawkey, 1987;

Table 2. Selected Typing Systems That May Be Useful
During Outbreaks of Nosocomial Infection[a]

---

1. Pattern of susceptibility to antimicrobial agents ("anti-
   biogram") or to heavy metals ("resistotyping")
2. Pattern of biochemical reactions ("biotyping")
3. Pattern of bacteriophage susceptibility ("phage typing")
4. Immunologic reactions ("serotyping")
5. Bacteriocin production or susceptibility (e.g., pyocin
   typing for Pseudomonas aeruginosa)
6. Plasmid profile or nucleic acid homology
7. Analysis of enzyme production
8. Analysis of marker proteins
9. Dienes reaction
10. Serum opacity
11. Colony morphology
12. RNA electrophoresis
13. Killer system analysis
14. Cytotoxicity assay

---

[a]Adapted from McGowan, 1985

McGowan, 1985; Sutherland, 1988), so that a large number of
potential methods for typing are now available (Table 2). On
occasion, use of more than one system provides very useful
epidemiologic information (Bacon et al., 1988). Most of the
methods listed are rarely performed on a routine basis, and
many are beyond the resources of the usual clinical laboratory.
However, a number can be conducted when circumstances dictate,
and others can be provided at reference laboratories. The laboratory director must be aware of the tests that are available,
must determine which can be performed on site when needed, and
must identify centers where other methods are available on
request.

Some of the newer systems (plasmid profiles, restriction
endonuclease analysis, chromosome analysis, etc.) examine the
genotype of the epidemic organism (Hawkey, 1987). These techniques make it possible to identify outbreaks that are not confined to specific species or organisms (Goldmann, 1987). For
example, the same plasmid carrying a gentamicin-resistance gene
has been found in three different multiresistant gram-negative
organisms, each isolated from widely separated centers in the
United States and in Venezuela (O'Brien et al., 1987). Thus,
an outbreak of resistant hospital organisms may be explained by
spread of a plasmid to several different organisms within the
hospital, rather than by transmission of resistant organisms of
a certain genus and species (Goldmann, 1987; O'Brien, 1987).
Fortunately, methods to enhance our knowledge of the spread of
resistance factors are becoming more widely available (Mayer,
1988; Schlaes and Currie-McCumber, 1986; Wachsmuth, 1986).
Molecular techniques are also beginning to define the epidemiology of nosocomial outbreaks of viral infection (Goldmann,
1987; Sutherland, 1988).

POTENTIAL PROBLEMS RELATED TO LABORATORY ACTIVITIES IN AN
OUTBREAK

1.   Misdiagnosis of Outbreak ("Pseudooutbreak")

On occasion, the existence of an epidemic is postulated
when none is actually present (Kusek, 1981). These episodes
occur when one has a clustering of apparent infections in which
infection is not present, or when the pattern of infections
give a false appearance of clustering.

Spurious outbreaks of nosocomial infection have been
traced to a variety of sources (Table 3). For example, at our
hospital a pseudooutbreak of Serratia marcescens bacteremia
finally was attributed to transfer of blood from unsterile
specimen containers to blood culture collection vials (Hoffman
et al., 1976). Another source of special importance to the
laboratory is inaccurate or inconsistent microbiologic
procedures. For example, a recent report of 23 cases of
Legionella infection in a single hospital apparently resulted
from use of a new probe technique for diagnosis of legionella;
the test had poor specificity, and many false-positive tests
were recorded (Laussucq et al., 1988). Pseudooutbreaks can
take a considerable amount of time to identify. On occasion
someone in the laboratory finds a problem in technique or
procedure that brings the pseudooutbreak to light; usually,
however, it is the clinician who recognizes the problem by
noting a great disparity between the patient's clinical status
and the laboratory results.

Table 3.   Sources of Pseudooutbreaks of Nosocomial Infection

---

A. Related to the clinical entity
   1. Wrong diagnosis of clinical entity
   2. Positive cultures represent colonization rather than
      infection
   3. Failure to distinguish community-acquired from
      nosocomial onset

B. Related to laboratory
   1. Contamination during specimen collection
   2. Contamination during specimen transport
   3. Contamination during processing
      a. Media
      b. Equipment
   4. Use of inadequate method or technique

C. Related to case-finding
   1. Increased surveillance efficiency
   2. Improved laboratory techniques for identification

D. Related to chance clustering

---

2. Overinterpretation of Typing Procedures

Typing procedures represent a great advance in the ability to characterize an outbreak. Nevertheless, there are certain problems associated with their use. Interpretation is a potential pitfall - on occasion, typing is interpreted to show that organisms are identical. In reality, however, these procedures rarely can prove that organisms are identical. Instead, they usually show only that organisms are not different. This considerable distinction focuses attention on the possibility that similar reactions can be obtained by chance alone.

3. Quality Control for Typing Procedures

Outbreak organisms rarely are all isolated at the same time. Thus, typing procedures often will be performed at different times. The need for control of batch-to-batch variation in media, phages, etc., is especially crucial with many of these typing procedures, and this explains why many of the procedures are not performed in the usual clinical laboratory. In some cases, one may need to go back and type isolates recovered at different time periods together on the same media, etc., to remove confounding variables of the testing process.

4. Needless Discrimination of Organism Differences

The use of too many individual markers in each system can cause as many problems as use of too few discriminators. It is important for the laboratory to avoid going overboard on use of markers to differentiate strains (Graham et al., 1985). For example, use of 20 different antibiotics, where one defines organisms as different if their pattern of susceptibility to these drugs varies by more than one drug, almost guarantees that organisms which actually are part of the outbreak will be identified as different and thus considered unrelated.

5. Overuse of Typing Techniques

One must be selective in the use of typing systems, even when an outbreak occurs. In some outbreaks, control measures can resolve the problem even without microbiologic data or epidemiologic investigation (Haley et al., 1985). In other situations, use of simple marker systems such as biotyping and antibiotic resistance pattern will sufficiently characterize the situation that no further testing is needed. For example, Harstein et al. (1987) found that plasmid DNA profile analysis is a more discriminating tool for analysis of multiple strains of coagulase-negative staphylococci, but added little to use of a disk diffusion antimicrobic susceptibility panel, which was sufficient to provide an understanding of the epidemiology of occurrence. Use of more complicated typing systems should be reserved for situations where control measures fail or where organisms or infection occurrence are being studied for academic reasons.

CONCLUSION

The laboratory may face exceptional demands for service at the beginning of, and throughout, an epidemic period (Goldmann and Macone, 1980). Yet, these activities are essential in the fight against outbreaks of hospital infection. The activities must be carried out in rapid and effective fashion, yet must be as efficient and cost-effective as possible. For all of these attributes to be blended, the microbiologist must prepare in advance for the various diagnostic tasks, whether performed on site or by referral sources. Contingency plans must be made for the types of outbreaks that have occurred most frequently in the past in a given hospital so that the laboratory is ready to deal with the exceptional requests in smooth fashion. After the outbreak occurs is too late to begin to deal with the laboratory response to an epidemic.

REFERENCES

Bacon, A.E., Fekety, R., Schaberg, D.R., and Faix, R.G., 1988, Epidemiology of Clostridium difficile colonization in newborns: results using a bacteriophage and bacteriocin typing system, J. Inf. Dis., 158:349-354.

Bartlett, C.L., 1987, Efficacy of different surveillance systems in detecting hospital-acquired infections, Chemioterapia, 6:152-155.

Coyle, M.B., 1984, DRGs and the practice of clinical microbiology, Clin. Microbiol. Newsletter, 6:175-179.

Dixon, R.E., 1986, Investigation of endemic and epidemic infections, in: "Hospital Infections," 2nd edition, J.V. Bennett and P.S. Brachman, eds., Little Brown and Company, Boston.

Farmer, J.J., III, 1988, Conventional typing methods, J. Hosp. Infect., 11(Suppl. A):309-314.

Goldmann, D.A., and Macone, A.B., 1980, A microbiologic approach to the investigation of bacterial nosocomial infection outbreaks, Infect. Control, 1:391-400.

Goldmann, D.A., 1987, New microbiologic techniques for hospital epidemiology, Eur. J. Clin. Microbiol., 6:344-347.

Graham, D.R., Dixon, R.E., Hughes, J.M., and Thornsberry, C., 1985, Disk diffusion antimicrobial testing for clinical and epidemiologic purposes, Am. J. Infect. Control, 13:241-249.

Haley, R.W., Tenney, J.H., Lindsey, J.O., II, Garner, J.S., and Bennett, J.V., 1985, How frequent are outbreaks of nosocomial infection in community hospital? Infect. Control, 6:233-236.

Hartstein, A.I., Valvano, M.A., Morthland, V.H., Fuchs, P.C., Potter, S.A., and Crosa, J.H., 1987, Antimicrobic susceptibility and plasmid profile analysis as identity tests for multiple blood isolates of coagulase-negative staphylococci, J. Clin. Microbiol., 25:589-593.

Hawkey, P.M., 1987, Molecular methods for the investigation of bacterial cross-infection, J. Hosp. Infect., 9:211-218.

Hoffman, P.C., Arnow, P.M., Goldmann, D.A., Parrott, P.L., Stamm, W.E., and McGowan, J.E., Jr., 1976, False-positive blood culture - association with nonsterile blood collection tubes, J.A.M.A., 236:2073-2075.

Hoffman, S.A., and Moellering, R.C., Jr., 1987, The
      enterococcus - "putting the bug in our ears," Ann.
      Intern. Med., 106:757-761.
Jacobson, J.T., 1985, The clinical microbiology laboratory and
      hospital infection control, Laboratory Medicine,
      16:243-247.
Kusek, J.W., 1981, Nosocomial pseudoepidemics and pseudoinfec-
      tions: an increasing problem, Am. J. Infect. Control,
      9:70-75.
Laussucq, S., Schuster, D., Alexander, W.J., Thacker, W.L.,
      Wilkinson, H.W., and Spika, J.S., 1988, False-positive
      DNA probe test for Legionella species associated with a
      cluster of respiratory illness, J. Clin. Microbiol.,
      26:1442-1444.
Mayer, L.W., 1988, Use of plasmid profiles in epidemiologic
      surveillance of disease outbreaks and in tracing the
      transmission of antibiotic resistance, Clin. Microbiol.
      Rev., 1:228-243.
McGowan, J.E., Jr., 1985, Role of the microbiology laboratory
      in prevention and control of nosocomial infection,
      in: "Manual of Clinical Microbiology," 4th Edition,
      E.H. Lennette, A. Balows, W.J. Hausler, Jr., and H.J.
      Shadomy, eds., American Society for Microbiology,
      Washington.
O'Brien, T.F., Members of Task Force 2, 1987, Resistance of
      bacteria to antibacterial agents: report of task force
      2, Rev. Inf. Dis., 9(Suppl. 3):S244-S260.
Parkhurst, S.M., Blaser, M.J., Laxson, L.B., and Wang, W.L.,
      1985, Surveillance for the detection of nosocomial
      infections and the potential for nosocomial outbreaks.
      II. Development of a laboratory-based system, Am. J.
      Infect. Control., 13:7-15.
Patterson, J.E., Colodny, S.M., and Zervos, M.J., 1988, Serious
      infection due to beta-lactamase producing Streptococcus
      faecalis with high-level resistance to gentamicin, J.
      Inf. Dis., 158:1144-1145.
Schlaes, D.M., and Currie-McCumber, C.A., 1986, Plasmid analysis
      in molecular epidemiology - a summary and future
      directions, Rev. Inf. Dis., 8:738-746.
Sutherland, S., 1988, Viral typing, J. Hosp. Infect., 11(Suppl.
      A):315-319.
Wachsmuth, K., 1986, Molecular epidemiology of bacterial
      infections: examples of methodology and of investiga-
      tions of outbreaks, Rev. Inf. Dis., 8:682-692.
Weigelt, J., Haley, R., and Siebert, G., 1987, Necessity and
      efficiency of wound infection surveillance after
      discharge, Am. J. Infect. Control, 16:75, 1988.
Zervos, M.J., Kauffman, C.A., Therasse, P.M., Bergman, A.G.,
      Mikesell, T.S., and Schaberg, D.R., 1987, Nosocomial
      infection by gentamicin-resistant Streptococcus
      faecalis - an epidemiologic study, Ann. Intern. Med.,
      106:687-691.

METHICILLIN-RESISTANT STAPHYLOCOCCUS AUREUS (MRSA) - IS IT CONTROLLABLE?

C. Glen Mayhall

Hospital Epidemiology Unit
Division of Infectious Diseases
Department of Medicine
Medical College of Virginia
Virginia Commonwealth University
Richmond, Va. 23298

INTRODUCTION

Methicillin was first used in 1959. Within 2 years, the first Staphylococcus aureus isolate resistant to methicillin was reported by Jevons (M.P. Jevons, letter to the Editor, Br. Med. J. 1:124-125, 1961). During the 1960s, methicillin-resistant S. aureus (MRSA) became a major problem in Europe only to recede in the early 1970s (30). Although outbreaks were reported from the United States in 1968 by Barrett et al. (7) and in 1970 by O'Toole et al. (42), MRSA did not become a problem in this country until the mid 1970s (25). The MRSA epidemic has continued to expand through the 1980s (28,45,66). In spite of intense efforts, attempts to control MRSA outbreaks in hospitals and to eradicate MRSA from hospital patient populations have either failed or met with very modest success. Even though MRSA has been difficult to control, much has been learned about the epidemiology of MRSA in the hospital over the last decade. Available data indicate that MRSA can be controlled but that it is difficult to eliminate MRSA from hospital patient populations. Design of control programs for MRSA require an understanding of the mechanisms of methicillin resistance in S. aureus, how to identify MRSA, the origin of MRSA in the hospital, the reservoirs, mode of transmission, risk factors for acquiring MRSA and the application of control measures that have proven effective in studies of outbreaks and their control.

MECHANISM OF RESISTANCE TO METHICILLIN

Although resistance to many antibiotics is plasmid mediated in MRSA, the genetic determinants for methicillin resistance are located in the chromosome. MRSA appears to develop under the selective pressure of potent broad spectrum antibiotics given to severely ill patients in tertiary care medical centers.

Methicillin resistance is also referred to as intrinsic resistance and is not due to production of beta-lactamases by MRSA. Rather, methicillin resistance is due to a change in proteins to which penicillin and other beta-lactam antibiotics bind in the cell membrane (26). These proteins are called penicillin-binding proteins or PBPs. PBPs are enzymes which are probably important in cell wall synthesis. The PBPs in methicillin-resistant S. aureus have a low affinity for methicillin and all other beta-lactam antibiotics.

Due to an unknown mechanism, most cells in a MRSA population do not express the resistant phenotype. Expression of resistance by a small proportion of cells in a strain is termed heteroresistance. The resistant cells grow slowly, and resistance may not become apparent in standard susceptibility tests unless incubation is extended beyond 24 hours. However, the expression of resistance may become obvious after a shorter incubation period if a lower incubation temperature (30°C) is used or if the osmolality of the growth medium is increased. Since MRSA may not be identified in standard susceptibility tests in hospital laboratories, special susceptibility tests for the reliable detection of MRSA must be used. Only if MRSA can be reliably identified in the laboratory can a control program for MRSA be applied effectively.

ORIGIN OF MRSA THAT CAUSE HOSPITAL OUTBREAKS

Methicillin-resistant S. aureus are probably generated in medical school-affiliated tertiary care centers by the treatment of very ill patients with broad spectrum antibiotics. However, the mechanism by which this occurs is unknown. Transmission of MRSA strains between hospitals and between nursing homes and hospitals has been well documented (9,10,27,35,42,44, 46,54,55,62,65). In a few localities patients admitted to hospitals may be infected with community-acquired MRSA (33,53). Another source may be patients who are readmitted after having become colonized and/or infected with MRSA during a previous admission to the same hospital (2,9,38,64,65). Thus, an unknown proportion of MRSA isolates may be generated in tertiary care hospitals, MRSA strains may be reintroduced into the hospital when patients are readmitted and patients colonized or infected with MRSA may be transferred from other institutions. In a few localities patients, particularly drug addicts, may be admitted with strains of MRSA acquired in the community.

HOSPITAL RESERVOIRS FOR MRSA

The major reservoirs for MRSA in the hospital are colonized or infected patients and personnel. MRSA has been reported to cause infections in patients at many sites. Types of infections that have been reported include bacteremia, intravascular catheter infection, endocarditis, phlebitis, wound infection, pneumonia, empyema, lung abscess, urinary tract infection, skin and soft tissue infection, intraabdominal abscess, peritonitis, enterocolitis, meningitis and conjunctivitis (2,3,8,9,11,15,17, 23,31,32,34,35,37,38,41-53,65). Patients may also become colonized with MRSA at several body sites including the nose, throat, axilla, umbilicus, wound surfaces, eczematous skin, perineum, and lower gastrointestinal tract and rectum (1,2,8,9,

34

17-19,40,42,44,46,50,64,65). Many patients with colonization never develop infection but may remain colonized for many months. The reservoir of colonized patients may be just as important a source of MRSA as are patients infected with MRSA, but colonized patients are more difficult to identify.

The second reservoir for MRSA is made up of colonized and infected hospital personnel. Personnel are more likely to become colonized than infected. Although there are many reports of colonization of health-care workers during outbreaks (3,5,9,11,15,19,29,31,34,35,39,44,46,57,65), in only a few reports could hospital personnel be implicated in transmission of MRSA to patients (29,31,35,65). Occasionally, hospital personnel become infected with MRSA, but there are no reports of infected personnel acting as a source for patients (9,16,19,29,54). Types of infections reported in personnel include infections of the hands (infections of cuts, eczematous eruptions and paronychia), forearms, and conjunctiva.

Although less important than patients and personnel, the environment may also be a source for MRSA that colonize and infect patients. Except for MRSA outbreaks in burn units, most outbreaks have been accompanied by minimal environmental con- tamination (9,11,15,31,39,42,49,58,65). However, there is some evidence that the environment may play a minor role as a source of MRSA. Simpson and coworkers identified silicone oil baths as a source of MRSA that caused infections in two of their patients (59). Berman and colleagues recovered MRSA from 7 of 24 tourniquets used by medical house staff and hospital phlebotomists (D.S. Berman, S. Schaefler, and M.S. Simberkoff, letter to the Editor, N. Engl. J. Med. 315:514-515, 1986). Most tourniquets were used on 15 to 20 patients a day and were used for periods of 3 weeks to 6 months. Bitar and coworkers carried out an extensive culture survey of the environment during a MRSA outbreak in their hospital (9). While only 16 cultures (1.26%) of 1268 cultures taken were positive for MRSA, they made two important observations. First, the environmental surfaces that were contaminated with MRSA were all surfaces in frequent contact with hands, such as telephones, bed crank, bed rail, over bed light, cabinet door handle, chart cover, plate stamping machine, refrigerator door handle and isolation cart. Second, since all of these cultures were taken after terminal cleaning, it was apparent that surfaces contaminated with MRSA were difficult to decontaminate.

In contrast to other areas of the hospital the environment of burn units may become heavily contaminated with MRSA during outbreaks. Crossley and associates found that 33 (22.8%) of 145 cultures taken in their burn unit during a MRSA outbreak were positive for MRSA (16). They also noted difficulty with decontaminating the surfaces of hydrotherapy tubs. Rutala and colleagues also noted extensive environmental contamination during a MRSA outbreak in their burn unit (52). Cultures of air from rooms of patients with MRSA infection gave mean counts of 1.9 CFU of MRSA per cubic foot. Mean counts for elevated surfaces and floors in these rooms were 20 and 48 CFU of MRSA per Rodac plate, respectively. Surface contamination was equally heavy in the hydrotherapy room. Other areas contiguous to patient rooms were also contaminated with MRSA but at lower levels. Although cleaning rooms after discharge of patients markedly lowered contamination with MRSA, the organism could still be recovered at low levels from air and surfaces. This observation is consistent with those of Bitar and coworkers who noted difficulty in eliminating MRSA from contaminated surfaces (9).

In summary the reservoir for MRSA in the hospital is made up by colonized and infected patients and colonized personnel. Except for outbreaks in burn units, the environment is usually not an important source for MRSA. However, when environmental contamination does occur, MRSA are usually found on surfaces touched frequently by hands, and MRSA may be difficult to remove from these surfaces.

MODE OF TRANSMISSION FOR MRSA

The great bulk of evidence supports direct contact as the principal mode of transmission for MRSA. Transmission probably occurs most often by contact with the contaminated hands of health-care workers. Like the evidence for hand transmission of other microorganisms in the hospital, the evidence for transmission of MRSA by hands of personnel is highly suggestive but not definitive (9,15,16,19,40,43,46). Crossley and associates recovered MRSA from the hands of nurses after they had changed the dressings of infected patients and before, but not after, they washed their hands (16). Craven and coworkers implicated a nurse with MRSA colonization of chronic dermatitis in spread of MRSA in their hospital (15). Other indirect evidence that hand transmission may take place is the observation of Bitar and coworkers that all of the environmental surfaces contaminated by MRSA were those surfaces frequently in contact with hands (9).

In addition to transmission between patients by the contaminated hands of personnel, health-care workers who are carriers of MRSA may transmit MRSA to patients by first contaminating their hands from their site of carriage (nose, perineum, site of chronic dermatitis) and then having hand contact with the patient. Evidence for this latter type of transmission was provided by Bitar and coworkers who observed that nasal and hand cultures taken simultaneously in three nasal carriers were all positive (9). Further, these nasal carriers were the only personnel among 303 physicians and nurses from who hand cultures were taken who had MRSA recovered from their hands.

Whether transmission occurs between contaminated environmental surfaces and patients by way of the hands of personnel is theoretically possible, but there is even less evidence for this type of transmission than there is for transmission from patients to hands of personnel to patients. However, there is some evidence that makes it likely that transmission occurs occasionally from environmental surfaces to patients by way of hands of health-care workers. First, MRSA has been recovered from surfaces frequently touched by personnel (3,9,15,16,64). Second, MRSA are known to survive on inanimate surfaces for up to 3 hours and probably much longer. Given the difficulty in decontaminating the environment, MRSA may occasionally be transmitted between environmental surfaces and patients by the hands of health-care workers.

Although airborne transmission is frequently considered to be one of the modes of transmission for MRSA in the hospital, there is little evidence that this is a significant route of transmission for MRSA. In three reported outbreaks air cultures were negative for MRSA (31,39,46). In two outbreaks a single colony of MRSA was recovered from air samples (9,22) and in one outbreak only one of seven air cultures was positive for MRSA (43). All of the air cultures that have recovered larger numbers of microorganisms from the air have been performed in burn units (3,11,16,52). Even in these studies, MRSA were not recovered from many of the air samples and concentrations of MRSA in air were low. No studies have been reported that offer proof that microorganisms found in the air caused colonization or infections in patients. Further, isolation wards or units with special air handling systems are no more effective in control of MRSA outbreaks than are such areas without special ventilation systems (Duckworth, personal communication, cited in 12,19).

RISK FACTORS FOR COLONIZATION AND INFECTION WITH MRSA

There are 9 published reports of outbreaks of MRSA where case-control studies were conducted to identify risk factors for colonization and infection with MRSA (9,11,17,32,35,43,49, 51,65). In six of these studies controls were selected from among patients infected with methicillin-sensitive S. aureus (MSSA) (11,17,32,35,43,49). In all but one (32) of these studies data were analyzed using only univariate tests. In two case-control studies patients who were culture negative for MRSA were used as controls (9,65). In these two studies data were also analyzed by multivariate analysis. In one study of surgical wound infections caused by MRSA, controls were selected from among patients with wound infections caused by microorganisms other than MRSA, and a multivariate technique was also used in the analysis of the data (51).

Risk factors for acquisition of MRSA are shown in Table 1. Six of the reported investigations of outbreaks performed case-control studies that used patients infected with MSSA as controls (11,17,32,35,43,49). Patients with MSSA infection may not be appropriate controls for such case-control studies. Many patients with MSSA infections may have become infected with a strain of MSSA with which they were colonized prior to admission. Such patients with MSSA colonization may be less susceptible to colonization with MRSA due to competition between the two strains for colonization of a given site. Further, only those case-control studies that used MSSA controls showed a significant relationship between administration of antibiotics and infection with MRSA. This apparent relationship may be an artifact of the effect of antibiotics on patients colonized with MSSA. Thus, perhaps only patients colonized with MSSA who did not receive antibiotics developed MSSA infections giving the appearance of a relationship between treatment with antibiotics and acquisition of MRSA. Those studies that used patients culture negative for MRSA as controls failed to identify antibiotic therapy as a risk factor for MRSA colonization or infection (9,65). Unfortunately, most case-control studies have not systematically studied invasive procedures and instrumentation as possible risk factors for colonization and infection with MRSA.

Table 1. Risk Factors for MRSA Colonization and Infection

Factor	References
Duration of hospitalization	17,65
Period between admission and first culture positive for MRSA	11,17,32,35,49
Treatment with antibiotics	11,17,32,35,43,49
Multiple antibiotics	11,17,35,43,49
Cephalosporins	35,43,49
Aminoglycosides	11,35,39
Age	65
Severity of underlying disease	9,32,49,65
Intravascular devices	11,32,49
Surgery	9,32,49
Multiple operations	51
Wound drain tubes	51
Wound manipulations	9
Respiratory tract instrumentation	9,11
Tracheostomy	9
Endotracheal intubation	32
Nasogastric tube	9
Mechanical ventilation	11,32
Foley catheter	32,49

CONTROL OF EPIDEMIC MRSA

Unlike recognition of many types of outbreaks, recognition of a MRSA outbreak is not usually defined by a rise in the rate of infection. The outbreak is most often recognized by the appearance of MRSA isolates in a hospital where such isolates had not previously been identified. This is an important point, because control or eradication may be possible only when the outbreak is detected early after appearance of only a few cases.

An outbreak can be detected by either surveillance of infections in patients or surveillance of microorganisms isolated and identified in the hospital microbiology laboratory. However, either type of surveillance will be successful only if the laboratory can reliably and consistently identify MRSA. All isolates of S. aureus should be tested specifically for methicillin resistance. Methods for identification of MRSA include some combination of increased salt content in the media, lowered incubation temperature (35° or 30°C) and a longer incubation period.

A heightened level of awareness must be maintained, particularly for surgery and burn services. MRSA may enter a hospital when a patient colonized or infected with MRSA is transferred from another hospital or when a patient such as a drug addict is admitted with community-acquired MRSA (33,53).

Determination of the Extent of the Outbreak

For the purpose of control, in epidemiologic terms, a case in a MRSA outbreak is defined as a patient with a site culture positive for MRSA which may be a source for another patient. Thus, the culture positive site may be either a site of infection or a site of colonization. There are three ways to identify culture positive patients in a MRSA epidemic. First, laboratory reports may be monitored for MRSA (2,9,61). Second, point prevalence culture surveys may be carried out periodically in high risk patient populations and in populations where MRSA cases have previously been identified (3,9,20,23,24,61,65). Third, contact tracing may be carried out to identify new cases among patients recently exposed to an unisolated case (9). This is the best method for identifying new cases, because it provides a systematic approach to identifying the population at greatest risk, i.e., those patients known to have been in close proximity to a patient who is culture positive for MRSA. Bitar and coworkers found this technique to be more effective for identifying new cases than either monitoring hospital laboratory reports or performing point prevalence culture surveys (9). Contact tracing is a particularly useful approach early in an outbreak when the number of culture positive patients is small and when identification of cases and application of control measures may lead to eradication of MRSA from the hospital.

Sites from which cultures should be taken for point prevalence culture surveys and contact tracing include nose and throat, sputum or respiratory secretions, the surface of open surgical wounds and decubiti, axillae, inguinal area, perineum, rectum, purulent drainage from any area, umbilicus in newborns and burn wound in burn patients. Cultures should be taken with swabs that do not contain antibacterial agents. It is unclear whether dry swabs or moistened swabs are most effective for recovery of MRSA from skin and mucosal surfaces. Swabs should be plated immediately or placed in a holding medium. Swabs should be streaked to agar plates containing a selective medium. One such medium that has proven effective is staphylococcus 110 agar containing 20 μ per ml of methicillin.

Colonization may be low level and difficult to detect in some sites (K.G. Kristinsson, P. Fenton, and P. Norman, letter to the Editor, Lancet 1:274-275, 1987). This can be overcome by taking multiple cultures over time. This may be particularly important in burn patients where multiple cultures from many different sites may be needed. Another approach to detect light colonization is use of broth enrichment where swabs are placed in salt enrichment broth after streaking the plates (36).

Reservoir Reduction

Once the patient reservoir has been defined, control efforts should be directed at reservoir reduction and interruption of transmission. Patients who are culture positive for MRSA should be discharged as soon as possible (2,12,21). There is no evidence that they provide a risk to the community or their families (4).

Patients may be treated systematically and/or topically with antimicrobial agents in an attempt to eradicate colonization. Ward and associates treated colonized patients with rifampin and trimethoprim-sulfamethoxazole combined with intranasal bacitracin ointment and hexachlorophene baths on the first two days of therapy (65). For nasal colonization, rifampin and bacitracin worked as well as the full regimen of rifampin, trimethoprim-sulfamethoxazole and bacitracin. The full regimen eliminated MRSA from 13 of 16 extranasal sites in patients. No adverse reactions to these drugs were noted. No resistance developed when the complete regimen was used. Ellison and colleagues also treated patients with rifampin and trimethoprim-sulfamethoxazole (21). However, they used double the doses of these medications given by Ward and associates, but they did not use topical bacitracin and hexachlorophene baths. This regimen eradicated MRSA carriage in only two thirds of their patients. Ellison and colleagues noted development of rifampin resistance when the drug therapy failed. They observed that failure of the regimen was more likely when there was a foreign body at the colonized site. These authors did not recognize any secondary cases after contact with a patient treated with this protocol. They suggested that even a transient decrease in MRSA colonization may contribute to the control of a MRSA outbreak.

Pearman and colleagues treated 20 patients colonized with MRSA with rifampin and sodium fusidate for 2 weeks, and 85% remained free of MRSA colonization at 10 weeks after therapy (44). Like Ellison and colleagues, they found that therapy for eradication of MRSA colonization may fail when there is a foreign body present at the colonized site. Smith and coworkers used clindamycin to clear MRSA from 2 patients colonized by a clindamycin-susceptible strain (60).

Mupirocin, formerly called pseudomonic acid, has proven to be an effective topical agent for treatment of nasal and skin colonization with MRSA (12,13,18,19). Minimum inhibitory concentrations for MRSA are very low, and it has been difficult to induce resistance in vitro by exposing MRSA to increasing concentrations of mupirocin. Mupirocin has proven effective in eradicating colonization from nose, eczematous skin and decubiti. Toxicity from absorption is unlikely since it is rapidly hydrolyzed in the systemic circulation and its inactive metabolite has a half-life of less than 30 minutes in plasma (14).

Another approach for eradication of MRSA from sites of colonization was introduced by Tyzack (63). He used a topical povidone-iodine regimen. This included bathing patients with povidone-iodine in their bath water, application of povidone-iodine ointment to the nares of all intensive care and isolation patients and application of povidone-iodine ointment or povidone-iodine gauze pads to all wounds and pressure sores. Visitors and staff associated with MRSA patients washed their hands using a povidone-iodine surgical scrub. Tyzack found this program to be highly effective with a marked reduction in MRSA infections and elimination of MRSA bacteremia.

The second reservoir that must be identified and reduced is that in hospital personnel. As noted above, personnel frequently become colonized with MRSA, but in only a minority of instances has it been possible to epidemiologically implicate them in transmission of MRSA to patients. During outbreaks, personnel may be colonized in their nares, throat, axillae, inguinal areas, perineum, rectum or in areas where there are skin eruptions (35). They may also develop infections due to MRSA on their hands and forearms (16,19). During outbreak investigations, personnel in the involved area should be examined for signs of dermatitis and infections on hands and forearms. Cultures should be taken from nares, throat, areas of dermatitis and any inflamed areas on hands and forearms. Personnel who are culture positive for MRSA should temporarily be removed from duty and offered treatment for their colonization or infection (18,29,46,65).

Antimicrobial agents that may be used for eradication of MRSA colonization in personnel include topical bacitracin (34), topical bacitracin combined with rifampin (65), topical vancomycin (9), rifampin and sodium fusidate (44), topical chlorhexidine cream (46), and topical mupirocin (13). The drug of first choice would appear now to be mupirocin given topically in a 5 day course. Mupirocin can also be used effectively on areas of colonized skin such as perineal skin. Most personnel will have colonization limited to the nares, and they may return to duty 24 hours after start of therapy. However, all personnel with colonized skin eruptions and infected lesions on their hands and forearms must remain away from patients until their lesions have been treated and cleared.

The environment may be a third source for MRSA that are transmitted to patients. This is most likely to be the case in burn units where, unlike the ordinary patient unit, the environment may be heavily contaminated with MRSA (16,52). Thus, environmental decontamination may play an important role in control of MRSA outbreaks in burn units.

Unlike burn units, cultures in other patient care areas yield a low percentage of positive cultures for MRSA. However, MRSA can survive for hours on inanimate surfaces, and as noted above, surfaces touched by hands of patients and personnel are frequently positive for MRSA. Thus, in areas outside of the burn unit, the focus of environmental control should be decontamination of surfaces frequently contacted by hands such as sphygmomanometers, stethoscopes, IV poles, cabinet door handles, etc.

Interruption of Transmission

Since there is no evidence for any mode of transmission other than contact transmission, the infection control strategy is one of raising barriers to prevent contact transmission of MRSA to patients. Although strict isolation has been used by several authors as one approach to interrupting transmission of MRSA during outbreaks (9,11,15,47,65), data on transmission do not support this practice. Ribner and colleagues studied isolation procedures used in control of MRSA (48). They compared strict isolation with more limited isolation appropriate for the site of infection or colonization. Strict isolation was no more effective at preventing spread of MRSA than were other more specific types of isolation.

Barrier techniques are aimed largely at prevention of hand transmission. Hands should be washed before and after each patient contact. While handwashing appears very important, there are no data to prove that washing with an antiseptic preparation is more effective than washing with plain soap in preventing transmission of MRSA. Gloves may be used to prevent hand transmission (12,46,48), but they must be changed between each patient, and hands should be washed after removing the gloves. When hands are not obviously soiled an antiseptic solution in a squeeze bottle may be applied to the hands as a rapid method of hand decontamination, particularly when sinks are few in number or in busy understaffed units where time for handwashing may be severely limited (4,9).

When working with patients with large amounts of wound drainage, large open infected or colonized wounds or with other extensive infected or colonized areas, gowns or plastic aprons may be needed to prevent contamination of clothing (4). There are no data to support the use of masks as part of a barrier program (12).

Modifications of standard barrier precautions have also been used for control of outbreaks due to MRSA. One such modification is that of cohort isolation (4,12,46). All colonized and infected patients are located in one area and are cared for by personnel who are assigned to these patients and who do not work with other patients. One advantage to this type of isolation is that breaks in barrier technique are not as likely to result in colonization or infection of new patients. An extension of cohort isolation is use of an isolation unit (6,56,57). These units appear to be most popular in England where investigators are of the opinion that airborne spread is an important mode of transmission. Therefore, these units may also have special air handling systems to prevent transmission of MRSA by way of the air. It is unclear whether these units are effective in containing the spread of MRSA because of the additional barriers they provide or because of the increased level of discipline in application of barrier techniques brought about by the awareness that one is working in a special isolation unit.

Another type of modified barrier technique is institution of tight restrictions on geographic movement of patients in the hospital such as transfer between nursing units or services (9).

A final consideration in interruption of transmission of MRSA is the need for a system to identify patients, who are MRSA carriers, on readmission to the hospital. After discharge, patients colonized with MRSA may remain colonized for many months or even years (4). If they are not identified immediately on readmission and placed in isolation, they may be an unrecognized source for transmission of MRSA to other patients. Flagging their medical records on discharge will permit immediate recognition of their MRSA carrier status on readmission (2,9,12,61).

CONTROL OF ENDEMIC MRSA

Methicillin-resistant S. aureus may continue to plague university hospitals much as did MSSA of phage type 80/81 in the 1950s until some sort of change apparently took place in the organism making it less competitive and it disappeared. Thus, although manageable, MRSA will probably not be eradicated from most tertiary care centers by even the best of control programs. Therefore, most centers will have to develop programs for control of endemic MRSA. Endemic MRSA may arise from continued generation of new strains, from transfer of patients who are colonized or infected with MRSA from other hospitals or nursing homes, or from admission of colonized or infected patients from the community. Many patients who are colonized, but not infected, with MRSA may not be recognized as potential sources of MRSA for colonization or infection of other patients. Thus, to keep track of the total patient reservoir for MRSA would require ongoing and extensive surveillance cultures.

Due to the long term commitment of resources needed for control of endemic MRSA, a control program should meet several criteria. First, the program should be as simple and efficient as possible. Second, it should utilize data generated by the hospital laboratory. Third, the hospital laboratory must use appropriate techniques to reliably identify all S. aureus isolates that are resistant to methicillin. Fourth, the program should be periodically reviewed for its effectiveness.

Unfortunately, there are few scientific data on programs to control endemic MRSA. Walsh and coworkers described a system for control of endemic MRSA which appeared to work well in their institution (64). The program had 6 elements. First, clinical cultures were monitored for MRSA isolates. Second, weekly surveillance cultures were taken from wounds and tracheostomy sites of all medical and surgical patients and from sputum in intensive care unit patients. Third, appropriate techniques were developed in the hospital laboratory for accurate identification of MRSA. Fourth, all patients with cultures positive for MRSA were placed on strict isolation. Fifth, a MRSA registry was maintained so that all patients from whom MRSA had been cultured could be identified and isolated immediately on readmission. Sixth, an intensive educational program was established. Since the implementation of this system, Walsh and coworkers have observed low rates of MRSA colonization and infection and a low infection to colonization ratio. A cost analysis also showed that their prospective microbiological surveillance program cost far less than the cost of controlling MRSA outbreaks.

At the Medical College of Virginia, we have developed a control program for endemic MRSA based entirely on clinical culture data. A special test pathway (Figure 1) has been established in our hospital laboratory for all S. aureus isolates in addition to the routine identification and susceptibility tests. This system is set up for immediate notification of the Hospital Epidemiology Unit even before the laboratory report becomes available on the patient's floor. Thus, on identification of a MRSA isolate, the laboratory notifies the Epidemiology Unit immediately with the patient's name, hospital number, date of culture, culture site and location in the hospital. The floor is immediately contacted by a nurse epidemiologist, and the nursing staff are instructed to place the patient on the appropriate type of isolation. A weekly log of all MRSA positive patients in the hospital is maintained in the Epidemiology Unit. Each new patient that appears on the weekly log is entered into the MRSA computer data base. A line

listing of all MRSA patients can be printed out for any time
period. All MRSA patients are flagged on the hospital information system. On readmission, the flag appears on the computer
screen in the Admitting Office, and the patient is placed on
the appropriate type of isolation at admission.

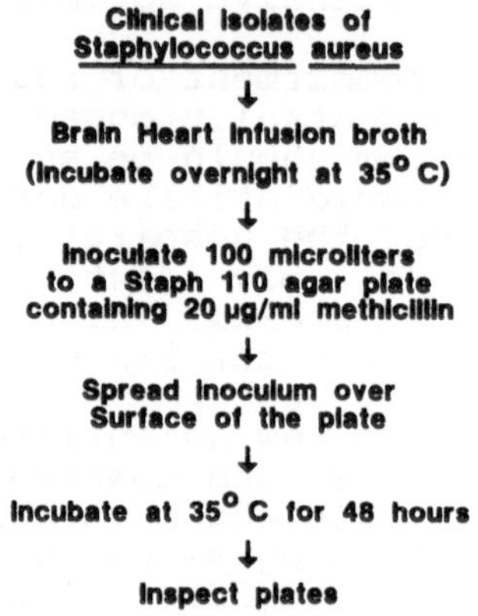

Clinical isolates of
Staphylococcus aureus
↓
Brain Heart Infusion broth
(Incubate overnight at 35°C)
↓
Inoculate 100 microliters
to a Staph 110 agar plate
containing 20 µg/ml methicillin
↓
Spread inoculum over
Surface of the plate
↓
Incubate at 35°C for 48 hours
↓
Inspect plates

Fig. 1. Procedure for detection of methicillin
resistance in isolates of S. aureus in the
Medical College of Virginia Hospital's
Microbiology Laboratory. All isolates of
S. aureus are tested with this procedure.

This control program has limited MRSA to the Surgery
Service, and the number of colonized and infected patients in
the hospital on any given day ranges between 10 and 20. MRSA
is occasionally isolated from patients on other services, but
most of these cases are in patients transferred from the
Surgery Service, and there is seldom evidence of transmission
of MRSA to other patients on the receiving service.

Although we have larger numbers of MRSA cases than Walsh
and coworkers, the effectiveness of our program may be comparable to theirs. Our institution is a tertiary care university
medical center with more than 3 times as many beds as the hospital of Walsh and coworkers. Our facility is a level one
trauma center, has 8 intensive care units and a 12 bed burn
center, has a large organ transplant program, has a cancer
center and provides care for many patients with the Acquired

Immunodeficiency Syndrome. Even though we do not do surveillance cultures, we identify a large number of colonized patients with data from clinical cultures. This is not surprising given the observation of Walsh and coworkers that surveillance cultures of wounds, tracheostomy sites and sputum were effective in identifying colonized patients. Clinical cultures are frequently taken from wounds and sputum and many of these patients turn out to be colonized rather than infected.

LITERATURE CITED

1. G.D. Aeilts, F.L. Sapico, H.N. Canawati, G.M. Malik, and J.Z. Montgomerie, Methicillin-resistant Staphylococcus aureus colonization and infection in a rehabilitation facility, J. Clin. Microbiol. 16:218-223 (1982).
2. S. Alvarez, C. Shell, K. Gage, J. Guarderas, D. Kasprzyk, J. Besing, and S.L. Berk, An outbreak of methicillin-resistant Staphylococcus aureus eradicated from a large teaching hospital, Am. J. Infect. Control 13:115-121 (1985).
3. P.M. Arnow, P.A. Allyn, E.M. Nichols, D.L. Hill, M. Pezzlo, and R.H. Bartlett, Control of methicillin-resistant Staphylococcus aureus in a burn unit: role of nurse staffing, J. Trauma 22:954-959 (1982).
4. G.A.J. Ayliffe, Guidelines for the control of epidemic methicillin-resistant Staphylococcus aureus, J. Hosp. Infect. 7:193-201 (1986).
5. A.E. Bacon, K.A. Jorgensen, K.H. Wilson, and C.A. Kauffman, Emergence of nosocomial methicillin-resistant Staphylococcus aureus and therapy of colonized personnel during a hospital-wide outbreak, Infect. Control 8:145-150 (1987).
6. B.A. Bannister, Management of patients with epidemic methicillin-resistant Staphylococcus aureus: experience at an infectious diseases unit, J. Hosp. Infect. 9:126-131 (1987).
7. F.F. Barrett, R.F. McGehee, Jr., and M. Finland, Methicillin-resistant Staphylococcus aureus at Boston City Hospital. Bacteriologic and epidemiologic observations, N. Engl. J. Med. 279:441-448 (1968).
8. C.A. Bartzokas, J.H. Paton, M.F. Gibson, R. Graham, G.A. McLoughlin, and R.S. Croton, Control and eradication of methicillin-resistant Staphylococcus aureus on a surgical unit, N. Engl. J. Med. 311:1422-1425 (1984).
9. C.M. Bitar, C.G. Mayhall, V.A. Lamb, T.J. Bradshaw, A.C. Spadora, and H.P. Dalton, Outbreak due to methicillin- and rifampin-resistant Staphylococcus aureus: epidemiology and eradication of the resistant strain from the hospital, Infect. Control 8:15-23 (1987).
10. B.V. Bock, K. Pasiecznik, and R.D. Meyer, Clinical and laboratory studies of nosocomial Staphylococcus aureus resistant to methicillin and aminoglycosides, Infect. Control 3:224-229 (1982).
11. J.M. Boyce, M. Landry, T.R. Deetz, and H.L. DuPont, Epidemiologic studies of an outbreak of nosocomial

methicillin-resistant <u>Staphylococcus</u> <u>aureus</u> infections, <u>Infect</u>. <u>Control</u> 2:110-116 (1981).

12. M.W. Casewell, Epidemiology and control of the "modern" methicillin-resistant <u>Staphylococcus</u> <u>aureus</u>, <u>J</u>. <u>Hosp</u>. <u>Infect</u>. 7(Suppl A):1-11 (1986).

13. M.W. Casewell, and R.L.R. Hill, The carrier state: methicillin-resistant <u>Staphylococcus</u> <u>aureus</u>, <u>J</u>. <u>Antimicrob</u>. <u>Chemother</u>. 18(Suppl A):1-12 (1986).

14. M.W. Casewell, and R.L.R. Hill, Mupirocin ("pseudomonic acid") - a promising new topical antimicrobial agent, <u>J</u>. <u>Antimicrob</u>. <u>Chemother</u>. 19:1-5 (1987).

15. D.E. Craven, C. Reed, N. Kollisch, A. DeMaria, D. Lichtenberg, K. Shen, and W.R. McCabe, A large outbreak of infections caused by a strain of <u>Staphylococcus</u> <u>aureus</u> resistant to oxacillin and aminoglycosides, <u>Am</u>. <u>J</u>. <u>Med</u>. 71:53-58 (1981).

16. K. Crossley, B. Landesman, and D. Zaske, An outbreak of infections caused by strains of <u>Staphylococcus</u> <u>aureus</u> resistant to methicillin and aminoglycosides. II. Epidemiologic studies, <u>J</u>. <u>Infect</u>. <u>Dis</u>. 139:280-287 (1979).

17. K. Crossley, D. Loesch, B. Landesman, K. Mead, M. Chern, and R. Strate, An outbreak of infections caused by strains of <u>Staphylococcus</u> <u>aureus</u> resistant to methicillin and aminoglycosides. I. Clinical studies, <u>J</u>. <u>Infect</u>. <u>Dis</u>. 139:273-279 (1979).

18. E.A. Davies, A.M. Emmerson, G.M. Hogg, M.F. Patterson, and M.D. Shields, An outbreak of infection with a methicillin-resistant <u>Staphylococcus</u> <u>aureus</u> in a special care baby unit: value of topical mupirocin and of traditional methods of infection control, <u>J</u>. <u>Hosp</u>. <u>Infect</u>. 10:120-128 (1987).

19. G.J. Duckworth, J.L.E. Lothian, and J.D. Williams, Methicillin-resistant <u>Staphylococcus</u> <u>aureus</u>: report of an outbreak in a London teaching hospital, <u>J</u>. <u>Hosp</u>. <u>Infect</u>. 11:1-15 (1988).

20. L.M. Dunkle, S.H. Naqvi, R. McCallum, and J.P. Lofgren, Eradication of epidemic methicillin-gentamicin-resistant <u>Staphylococcus</u> <u>aureus</u> in an intensive care nursery, <u>Am</u>. <u>J</u>. <u>Med</u>. 70:455-458 (1981).

21. R.T. Ellison, III, F.N. Judson, L.C. Peterson, D.L. Cohn, and J.M. Ehret, Oral rifampin and trimethoprim/sulfamethoxazole therapy in asymptomatic carriers of methicillin-resistant <u>Staphylococcus</u> <u>aureus</u> infections, <u>West</u>. <u>J</u>. <u>Med</u>. 140:735-740 (1984).

22. E.D. Everett, A.E. Rahm, Jr., T.R. McNitt, D.L. Stevens, and H.E. Peterson, Epidemiologic investigation of methicillin-resistant <u>Staphylococcus</u> <u>aureus</u> in a burn unit, <u>Milit</u>. <u>Med</u>. 143:165-167 (1978).

23. G.L. Gilbert, V. Asche, A.S. Hewstone, and J.L. Mathiesen, Methicillin-resistant <u>Staphylococcus</u> <u>aureus</u> in neonatal nurseries, <u>Med</u>. <u>J</u>. <u>Aust</u>. 1:455-459 (1982).

24. D.R. Graham, A. Correa-Villasenor, R.L. Anderson, J.H. Vollman, and W.B. Baine, Epidemic neonatal gentamicin-methicillin-resistant <u>Staphylococcus</u> <u>aureus</u> infection associated with nonspecific topical use of gentamicin, <u>J</u>. <u>Pediatr</u>. 97:972-978 (1980).

25. R.W. Haley, A.W. Hightower, R.F. Khabbaz, C. Thornsberry, W.J. Martone, J.R. Allen, and J.M. Hughes, The emergence of methicillin-resistant <u>Staphylococcus</u> <u>aureus</u> infections in United States hospitals. Possible role of the house staff-patient transfer circuit, <u>Ann</u>. <u>Intern</u>. <u>Med</u>. 97:297-308 (1982).

26. B.J. Hartman, and A. Tomasz, Expression of methicillin resistance in heterogenous strains of Staphylococcus aureus, Antimicrob. Agents Chemother. 29:85-92 (1986).

27. M.W. Humble, Imported methicillin-resistant Staphylococcus aureus infection: a case report, N.Z. Med. J. 84:476-478 (1976).

28. W.R. Jarvis, C. Thornsberry, J. Boyce, and J.M. Hughes, Pediatr. Infect. Dis. 4:651-655 (1985).

29. M.R. Jones, Outbreak of methicillin-resistant Staphylococcus aureus infection in a New Zealand hospital, N.Z. Med. J. 100:369-373 (1987).

30. C.T. Keane, and M.T. Cafferkey, Re-emergence of methicillin-resistant Staphylococcus aureus causing severe infection, J. Infect. 9:6-16 (1984).

31. J.J. Klimek, F.J. Marsik, R.C. Bartlett, B. Weir, P. Shea, and R. Quintiliani, Clinical, epidemiologic and bacteriologic observations of an outbreak of methicillin-resistant Staphylococcus aureus at a large community hospital, Am. J. Med. 61:340-345 (1976).

32. J.R. Lentino, H. Hennein, S. Krause, S. Pappas, G. Fuller, D. Schaaff, and M.B. DiCostanzo, A comparison of pneumonia caused by gentamicin, methicillin-resistant and gentamicin, methicillin-sensitive Staphylococcus aureus: epidemiologic and clinical studies, Infect. Control 6:267-272 (1985).

33. D.P. Levine, R.D. Cushing, J. Jui, and W.J. Brown, Community-acquired methicillin-resistant Staphylococcus aureus endocarditis in the Detroit Medical Center, Ann. Intern. Med. 97:330-338 (1982).

34. C.C. Linnemann, Jr., M. Mason, P. Moore, T.R. Korfhagen, and J.L. Staneck, Methicillin-resistant Staphylococcus aureus: experience in a general hospital over four years, Am. J. Epidemiol. 115:941-950 (1982).

35. R.M. Locksley, M.L. Cohen, T.C. Quinn, L.S. Tompkins, M.B. Coyle, J.M. Kirihara, and G.W. Counts, Multiply antibiotic-resistant Staphylococcus aureus: introduction, transmission and evolution of nosocomial infection, Ann. Intern. Med. 97:317-324 (1982).

36. R.R. Marples, and E.M. Cooke, Workshop on methicillin-resistant Staphylococcus aureus held at the headquarters of the Public Health Laboratory Service on 8 January 1985, J. Hosp. Infect. 6:342-348 (1985).

37. M. McDonald, A. Hurse, and K.N. Sim, Methicillin-resistant Staphylococcus aureus bacteraemia, Med. J. Aust. 2:191-194 (1981).

38. J.J. McNeil, A.D. Proudfoot, F.A. Tosolini, P. Morris, J.M. Booth, A.E. Doyle, and W.J. Louis, Methicillin-resistant Staphylococcus aureus in an Australian teaching hospital, J. Hosp. Infect. 5:18-28 (1984).

39. J.A.G. Melo Cristino, A.T. Pereira, F. Afonso, and J. Naidoo, Methicillin-resistant Staphylococcus aureus: a 6-month survey in a Lisbon paediatric hospital, J. Hyg. 97:265-272 (1986).

40. M.R. Millar, N. Keyworth, C. Lincoln, B. King, and P. Congdon, "Methicillin-resistant" Staphylococcus aureus in a regional neonatology unit, J. Hosp. Infect. 10:187-197 (1987).

41. J.P. Myers, and C.C. Linnemann, Jr., Bacteremia due to methicillin-resistant Staphylococcus aureus, J. Infect. Dis. 145:532-536 (1982).

42. R.D. O'Toole, W.L. Drew, B.J. Dahlgren, and H.N. Beaty, An outbreak of methicillin-resistant Staphylococcus aureus

infection. Observations in hospital and nursing home,
J.A.M.A. 213:257-263 (1970).

43. J.E. Peacock, F.J. Marsik, and R.P. Wenzel, Methicillin-resistant Staphylococcus aureus: introduction and spread within a hospital, Ann. Intern. Med. 93:526-532 (1980).

44. J.W. Pearman, K.J. Christiansen, D.I. Annear, C.S. Goodwin, C. Metcalf, F.P. Donovan, K.L. Macey, L.D. Bassette, I.M. Powell, J.M. Green, W.E. Harper, and M.S. McKelvie, Control of methicillin-resistant Staphylococcus aureus (MRSA) in an Australian metropolitan teaching hospital complex, Med. J. Aust. 142:103-108 (1985).

45. L.C. Preheim, D. Rimland, and M.J. Bittner, Methicillin-resistant Staphylococcus aureus in Veterans Administration Medical Centers, Infect. Control 8:191-194 (1987).

46. E.H. Price, A. Brain, and J.A.S. Dickson, An outbreak of infection with a gentamicin- and methicillin-resistant Staphylococcus aureus in a neonatal unit, J. Hosp. Infect. 1:221-228 (1980).

47. E. Rhinehart, D.M. Shlaes, T.F. Keys, J. Serkey, B. Kirkley, C. Kim, C.A. Currie-McCumber, and G. Hall, Nosocomial clonal dissemination of methicillin-resistant Staphylococcus aureus. Elucidation by plasmid analysis, Arch. Intern. Med. 147:521-524 (1987).

48. B.S. Ribner, M.N. Landry, and G.L. Gholson, Strict versus modified isolation for prevention of nosocomial transmission of methicillin-resistant Staphylococcus aureus, Infect. Control 7:317-320 (1986).

49. D. Rimland, Nosocomial infections with methicillin and tobramycin resistant Staphylococcus aureus - implication of physiotherapy in hospital-wide dissemination, Am. J. Med. Sci. 290:91-97 (1985).

50. D. Rimland, and B. Roberson, Gastrointestinal carriage of methicillin-resistant Staphylococcus aureus, J. Clin. Microbiol. 24:137-138 (1986).

51. H. Ross, Postoperative wound infection with methicillin-resistant staphylococci in general surgical patients, Aust. N.Z. J. Surg. 55:13-17 (1985).

52. W.A. Rutala, E.B.S. Katz, R.J. Sherertz, and F.A. Sarubbi, Jr., Environmental study of a methicillin-resistant Staphylococcus aureus epidemic in a burn unit, J. Clin. Microbiol. 18:683-688 (1983).

53. L.D. Saravolatz, D.J. Pohlod, and L.M. Arking, Community-acquired methicillin-resistant Staphylococcus aureus: a new source for nosocomial outbreaks, Ann. Intern. Med. 97:325-329 (1982).

54. G. Saroglou, M. Cromer, and A.L. Bisno, Methicillin-resistant Staphylococcus aureus: interstate spread of nosocomial infections with emergence of gentamicin-methicillin resistant strains, Infect. Control 1:81-89 (1980).

55. S. Schaefler, D. Jones, W. Perry, T. Baradet, E. Mayr, and C. Rampersad, Methicillin-resistant Staphylococcus aureus strains in New York City hospitals: inter-hospital spread of resistant strains of type 88, J. Clin. Microbiol. 20:536-538 (1984).

56. J.B. Selkon, E.R. Stokes, and H.R. Ingham, The role of an isolation unit in the control of hospital infection with methicillin-resistant staphylococci, J. Hosp. Infect. 1:41-46 (1980).

57. D.C. Shanson, D. Johnstone, and J. Midgley, Control of a hospital outbreak of methicillin-resistant

Staphylococcus aureus infections: value of an isolation unit, J. Hosp. Infect. 6:285-292 (1985).

58. D.C. Shanson, J.G. Kensit, and R. Duke, Outbreak of hospital infection with a strain of Staphylococcus aureus resistant to gentamicin and methicillin, Lancet 2:1347-1348 (1976).

59. C.N. Simpson, J. Ashford, and G. Duckworth, Silicone oil as a reservoir for nosocomial infection, J. Hosp. Infect. 10:91-94 (1987).

60. S.M. Smith, A. Mangia, R.H.K. Eng, P. Ruggeri, A. Cytryn, and F. Tecson-Tumang, Clindamycin for colonizataion and infection by methicillin-resistant Staphylococcus aureus, Infection 16:95-97 (1988).

61. R.P. Thompson, I. Cabezudo, and R.P. Wenzel, Epidemiology of nosocomial infections caused by methicillin-resistant Staphylococcus aureus, Ann. Intern. Med. 97:309-317 (1982).

62. D.E. Townsend, N. Ashdown, S. Bolton, J. Bradley, G. Duckworth, E.C. Moorhouse, and W.B. Grubb, The international spread of methicillin-resistant Staphylococcus aureus, J. Hosp. Infect. 9:60-71 (1987).

63. R. Tyzack, The management of methicillin-resistant Staphylococcus aureus in a major hospital, J. Hosp. Infect. 6(Suppl):195-199 (1985).

64. T.J. Walsh, D. Vlahov, S.L. Hanson, E. Sonnenberg, R. Khabbaz, T. Gadacz, and H.C. Standiford, Prospective microbiologic surveillance in control of nosocomial methicillin-resistant Staphylococcus aureus, Infect. Control 8:7-14 (1987).

65. T.T. Ward, R.E. Winn, A.I. Hartstein, and D.L. Sewell, Observations relating to an inter-hospital outbreak of methicillin-resistant Staphylococcus aureus: role of antimicrobial therapy in infection control, Infect. Control 2:453-459 (1981).

66. R.P. Wenzel, The emergence of methicillin-resistant Staphylococcus aureus, Ann. Intern. Med. 97:440-442 (1982).

INFECTIONS AMONG RESIDENTS OF NURSING HOMES: LESSONS FROM WISCONSIN AND THE RECENT LITERATURE

William E. Scheckler

Department of Family Medicine and Practice
and Department of Medicine
University of Wisconsin Medical School
Madison, WI

INTRODUCTION

Any discussion of infections occurring among residents of nursing homes and strategies for control of those infections needs to pay attention to several important demographic factors. Population trends clearly identify a major increase in the proportion of the population aged 65 and over developing over the next 50 years, and a substantial decrease in the percentage of the population under 25 years. The current ratio of 150 females per 100 males in the 65 and over age group is expected to continue (1). There is no reason to expect that the current proportion of 5% of the 65 and over population in nursing homes will decrease, and every reason to believe it will increase as the population over 80 years of age increases. Based on these figures, about 1,750,000 people 65 and over will be residing in nursing homes in the year 2000, a 300,000 increase from today.

To date there have been no national surveys or national studies of infections among residents of nursing homes. A widely cited statistic from a CDC paper (2) reporting on nosocomial infections in acute care hospitals is based on a few non-representative prevalence surveys of infections in nursing homes, and that statistic should probably be ignored. A number of statewide surveys of infection control practices and prevalence of infection have been done (3,4,5) and these are beginning to provide a picture of the issues of importance. Very few studies of nursing home associated infections have developed incidence data from ongoing surveillance over time (6,7). Most investigations have focused on individual problems such as urinary tract infections (8,9,10), pneumonia and influenza (11,12,13,14,15,16), and antibiotic use (17,18). The purpose of this report is to review the data developed in some comprehensive studies in eight rural Wisconsin nursing homes and compare the results with the increasingly available data from the literature of the last two years. Suggestions for infection control programs in nursing homes will then be presented based on these data.

In a prior study (6), infections occurring among residents in nursing homes were separated into three categories:

1. **Community acquired infections** - those developed when living in the community or at home.

2. **Nosocomial infections** - those acquired when residing in an acute care hospital.

3. **Nursing home associated infections** - the vast majority of all infections seen in nursing home residents - those acquired while a resident is in a nursing home.

The separation of infections into these three categories helps to better define their epidemiology and potential prevention strategies. Since there are substantial differences between nursing homes and acute care hospitals, reserving the term nosocomial infections for the acute care hospitals also helps clarify and focus analysis of these problems.

Residents in nursing homes differ from patients in acute care hospitals in a variety of ways. The setting of the nursing home is residential, long term care and terminal hospice care are more common, and the proportion of patients in "skilled care," "step down units" or rehab care is usually relatively small. This latter category is now more variable with the advent of "DRGs" and earlier discharges of elderly patients from hospitals. Residents of nursing homes are subject to reduced numbers of interventions, both diagnostic or therapeutic, and have much less in the way of acute disease or important acute exacerbations of chronic disease than do their counterparts in acute care hospitals. These important differences argue for a strategy that does not assume that everything important and useful in an acute care hospital infection control program can automatically be assumed to be appropriate for a nursing home. Some programs helpful in a nursing home population may also not be relevant to a hospital population.

The majority of information presented here was published in 1986 after a three year study in eight rural hospital-associated Wisconsin nursing homes (6). That study replicated Garibaldi's earlier prevalence survey in Utah (19), then expanded the methodology by performing sequential twice monthly prevalence surveys for six months in each home followed by six months of comprehensive ongoing surveillance data concerning infections among the residents in each home. In addition, early in 1988 the nursing homes were resurveyed by mailed questionnaire and telephone to collect 1987 surveillance data and follow-up information concerning the impact of the earlier study on their infection control practices. This Wisconsin data will then be compared and contrasted with studies from the current literature in the discussion section. The conclusion will present an infection control strategy appropriate to most nursing homes.

THE RURAL WISCONSIN DATA

Methods

The methods for conducting the prevalence and surveillance

surveys and other data collection in the eight rural Wisconsin nursing homes were detailed in a previous report (6). The infection control practitioners in those homes continued to do the same type of ongoing infection surveillance in 1987. By means of mailed questionnaires and telephone discussions in 1988, the 1987 surveillance data and other relevant aspects of the nursing home infection control program was obtained and reviewed. In 1986 two of the rural hospitals, each of which had an attached nursing home, consolidated, locating all hospital functions at one site and all nursing home functions at the other site. Therefore, the number of nursing homes in 1987 was reduced from eight to seven, but the population served and the total number of residents in the homes was about the same.

Profiles of the Residents

An analysis of all resident charts was done at the initial 1984 prevalence survey to describe the population in detail. Such descriptions allow those working in other nursing homes to look for similarities in population profile to their own home. The assumption here is that the more similarities in population the more likely the problems and issues concerning infection control will be the same. Table 1 and Table 2 provide a clear picture of the residents of the nursing homes in 1984. Current information suggests that profile is still accurate in 1988. In summary, the nursing home population is quite elderly, largely female, and each resident has usually three significant chronic diseases, at least one of which is the major reason why they cannot continue to live in an independent or minimal support environment. So far, the issue of debilitated patients with AIDS requiring nursing home care in these rural Wisconsin nursing homes is virtually nonexistent.

The nursing homes varied in size from 18 to 102 beds with occupancy from 82% to 100% in 1984. The payment source for the 403 residents was 63.3% Medicaid, 32.5% private pay and 4.5% Medicare. By state Medicaid definitions, two thirds of the residents were receiving skilled care and one third were receiving intermediate care. These proportions have stayed fairly constant in the 1980s with the state Medicaid Program being the major source of funding for most nursing home residents. Because of this, state and federal rules about eligibility for Medicaid funding and amount of reimbursement for nursing home care are crucial issues both for the nursing home resident, potential resident and for the nursing home itself.

Table 3 presents the infections defined in the initial 1984 prevalence survey and compares them to a similar survey in urban Salt Lake City reported in 1981 by Garibaldi and his colleagues. All three types of infections: nursing home associated, nosocomial and community acquired are included in the survey since the Utah group did not separate their infections by the three categories. Based on the Wisconsin analyses, however, it is reasonable to assume that about 90-95% of all three infections are in the nursing home associated category. Of interest in the Wisconsin data (6) is that presence of infected decubitus ulcers was statistically significantly associated with the presence of fecal incontinence. Presence of active infection was not associated with age or length of stay.

Table 1. Profile of 403 Residents - January to February 1984:
Prevalence Survey in Eight Rural Wisconsin Nursing Homes[a]

Characteristics of Patients	Findings
Age, years	Average, 83.4
	22 < 65
	119 ≥ 90
Marital status	
Married	54 (13.4)
Single	60 (14.9)
Widowed	275 (68.2)
Divorced	14 (3.5)
Gender	
Female	306 (75.9)
Male	97 (24.1)
Clinical status	
Stable	336 (83.4)
Deteriorating	58 (14.4)
Terminal	9 (2.2)
Ambulatory status	
Ambulatory	205 (50.9)
Bed/chair confined	158 (39.2)
Nonambulatory	40 (9.9)
Average length of stay at time of survey	42 mo.

[a]Number in parentheses indicates percent of patients.
From ref. 6: Arch. Intern. Med., 146:1981-1984 (1986).

Table 2. Rural Wisconsin Nursing Home Residents:
Prevalence of Active Diseases in 403 Residents

Diseases[a]	Number	Percent
Cardiovascular Diseases		
1. Organic heart disease	232	(57.6)
2. Hypertension	68	(16.9)
3. Peripheral vascular disease	67	(16.6)
CNS Diseases		
1. Organic brain syndrome	163	(40.4)
2. Cerebrovascular accident	103	(25.6)
3. Psychiatric disease	53	(13.2)
4. Other CNS disease	54	(13.4)
5. Mental retardation	10	(2.5)
Musculoskeletal Diseases		
1. Arthritis	153	(38.0)
2. Fracture	88	(21.8)
Other Diseases		
1. Diabetes mellitus	71	(17.6)
2. Cancer	46	(11.4)
3. Chronic obstructive pulmonary disease	45	(11.2)
4. All other diseases	181	(44.9)

[a]Average of 3.4, median of 3 diseases per resident.
From ref. 6: Arch. Intern. Med., 146:1981-1984 (1986).

Table 4 presents the infections defined in the rural
Wisconsin nursing homes by surveillance in the six month studies
done in 1984 and 1985 (6), during the follow-up period in 1987
and in combined data from two Maryland nursing homes (7). The
data are all expressed as incidence of infections calculated as
number of infections occurring over "100 resident months." It
is of interest that the rural Wisconsin and urban Maryland data
for the 1984 time period are so similar - 10.7 infections/100
resident months in Wisconsin versus 11.1 infections/100 resident
months in Maryland. The 1987 Wisconsin data shows an increase
to 12.6 infections/100 resident months which may reflect differences in reporting, differences in resident population or a
true increase over the earlier period.

Urinary tract infections are usually identified in both
prevalence and incidence studies as being among the most common
infections identified. In the rural Wisconsin prevalence survey
(6), all residents with an indwelling foley catheter (26 residents) or a suprapubic catheter (3 residents) in place had urine
specimens obtained in a standard sterile fashion for culture.
In 28 of the 29 specimens, 10^4 or more organisms were recovered,
the majority revealing 2 or more pathogens per urine specimen.
In the Utah study (19) a few years earlier, very similar findings were noted, 45 of 53 urine specimens collected from residents with indwelling catheters had asymptomatic bacteriuria
defined by 10^5 colony forming units per milliliter of urine and
81 bacterial isolates were recovered.

Table 3. Prevalence of Infections Among Nursing Home Residents[a]

Infection	1984 Wisconsin Survey of 403 Residents, No. of Infections (% Prevalence)	1981 Utah Survey (19) of 532 Residents, No. of Infections (% Prevalence)
Infected ducubitus ulcers	16 (4.0)	32 (6.0)
Conjunctivitis	5 (1.2)	18 (3.4)
Symptomatic urinary tract infection	14 (3.5)	14 (2.6)
Lower respiratory tract infection	10 (2.5)	11 (2.1)
Upper respiratory tract infection	2 (0.5)	8 (1.5)
Diarrhea; gastroenteritis	2 (0.5)	7 (1.3)
Bone	1 (0.3)	7 (1.3)
Other cutaneous	6 (1.5)	–
Total	56 (13.9)	97 (18.2)

[a]In the 1984 Wisconsin survey, there were 56 (13.9%) infections in 52 (12.9%) residents; in the 1981 Utah survey, there were 97 (18.2%) infections in 86 (16.2%) residents.
From ref. 6: Arch. Intern. Med., 146:1981-1984 (1986).

Table 4. Prospective Surveillance Based Incidence[a] of Infections Among Nursing Homes, Maryland (7) and Wisconsin (6)

Site of Infection	Rural Wisconsin 1984-85	1987	Maryland Homes "A & B" 1983-84
Urinary tract	4.8	5.1	3.8
Respiratory tract	3.0	3.9	3.4
Cutaneous - soft tissue	1.6	2.0	1.1
Eye - conjunctivitis	0.5	0.9	0.8
Gastroenteritis	0.6	0.9	1.4
Other	0.6	0.4	0.7
Total, all sites	10.7[b]	12.6[c]	11.1[d]

[a]Incidence expressed as a ratio of number of infections per 100 resident months.
[b]Represents 265 infections over 2479 resident months x 100 - 8 nursing homes.
[c]Represents 411 infections over 3267 resident months x 100 - 4 nursing homes (3 homes with 12 months of data, 1 home with 13 months of data).
[d]Represents 896 infections over 8068 resident months x 100 - 2 nursing homes.

It is not surprising that urinary tract infections are identified during surveillance with frequency since urine is generally easy to obtain and to culture. As will be noted in the discussion, the problem comes in assessing the clinical importance of the presence of bacteriuria in the elderly nursing home resident with or without a catheter and avoiding unnecessary antibiotic therapy.

The infection control programs in the rural Wisconsin nursing homes were better developed than reports from surveys in Minnesota (4), North Carolina (3) and Maryland (5) at the onset of the study in 1984. All of the Wisconsin homes had functioning infection control committees, an infection control practitioner in charge of the infection control program and standard policies for infection control. The study ended in mid 1986. For this reason all of the study nursing homes were resurveyed in 1988 to learn if the study had had an impact on their infection control program and what their program currently was doing. The seven remaining nursing homes (two merged) all continue to do ongoing surveillance. All have developed infection control manuals as a result of the study, incorporating policies and procedures shared by the participating homes; and 5 of the 7 note definite improvements and changes in their policies as a result of the study. For five of the seven homes, the ongoing relationships established by encouraging infection control practitioner membership in the regional Association for Practitioners in Infection Control Chapter has provided a network of colleagues for continuing consultation about shared problems. Most of these nursing homes did not have membership in APIC prior to the study as was also true of their parent nonprofit rural hospital. The opportunity to facilitate this network development was probably the single most important, lasting (so far) contribution of the original study.

DISCUSSION

As reported in the earlier paper (6), several features of the study in the rural Wisconsin nursing homes contrast with much of the current literature: there were very few clusters of infections, there were no multiply resistant gram negative rods and no methicillin resistant Staphylococcus aureus - therefore, no bad microbe reservoir for hospitals to worry about, there was minimal mortality directly attributable to infection, there were good staffing levels and minimal turnover of personnel compared to urban Utah (19), and finally, the accessibility to the parent rural hospitals probably facilitated ease of diagnostic procedures and more convenient accessibility of medical care including transfer to a hospital bed. Because of these differences, one must be very cautious about extrapolating the Wisconsin findings to nursing homes around the country, even in rural areas.

Certain patterns of similarity are beginning to become clear for most nursing homes, both urban and rural, however. The vast majority of the population of these homes (VA facilities excepted, of course) are women in a ratio of 2:1 to 3:1 over men - similar to their ratio in the overall over 65 population. The majority live as residents in the home for longer than two to three years. The patterns of infections seen by prevalence survey and incidence studies are more similar in site of infection and overall frequency than they are dissimilar.

The important issue remains, though, of what infections can and should be prevented and how this can be accomplished.

Urinary tract infections pose several substantial dilemmas. The Philadelphia group (20) has followed a large cohort of elderly for some time and reports a striking prevalence of bacteriuria which increases with age, with time, and in general does not seem responsible for either classic or general signs or symptoms of infection. Nicolle's controlled study (9) in an institutional elderly population of women has shown that antibiotic therapy has no benefit and some potential harm when used to treat asymptomatic bacteriuria identified by routine monthly cultures in patients without indwelling catheters. For catheterized residents of long term care facilities Kunin (21) recently showed in a prospective study that increased mortality was probably caused by associated medical conditions although catheterized residents did have more episodes of urinary tract infection and fever than noncatheterized residents. From the Wisconsin and other studies it is clear that bacteriuria is virtually always present in chronically catheterized nursing home residents and that the prevalence of bacteriuria will increase with age, especially in women, especially over 80 years of age. If we were to apply the Philadelphia data to nursing home resident populations without catheters and assume all catheterized residents had bacteriuria, both the prevalence and incidence of urinary tract infections would increase substantially from the data reported here and elsewhere. The conclusions seem clear from the foregoing:

1. Avoid chronic catheterization whenever possible.

2. Do not treat asymptomatic bacteriuria in the catheterized or noncatheterized resident.

3. Treat only those residents with clearcut infection or fever due to urinary tract bacteriuria after excluding other sources of fever.

Lower respiratory infections offer a number of options from the literature for prevention. Gross and his colleagues have demonstrated the efficacy of influenza vaccine in reducing mortality from an influenza outbreak (22). Others have also found clearcut efficacy of influenza immunization in this population (13,15,16). In a study by Marrie and colleagues from Canada (11), nursing home residents with nursing home acquired pneumonia admitted to a hospital were age and sex matched to individuals with community acquired pneumonia also admitted to a hospital. The nursing home residents with pneumonia had more dementia, cerebrovascular accidents and aspiration pneumonia than did their community controls. Although this study did not offer any suggestions for prevention, it did highlight the finding that as many as 1/4 to 1/3 of elderly patients with pneumonia may not have fever or may not have cough, whereas as many as 1/3 to 1/2 will have increasing confusion. For bacterial pneumonia, avoiding medication such as long acting benzodiazapenes that decrease levels of consciousness and could therefore facilitate aspiration pneumonia seems reasonable. Being alert to swallowing difficulties, especially in patients with a stroke, is also an important aspiration avoidance strategy. Finally, a one time dose of pneumovax is probably reasonable and is recommended by the CDC (23) although controversy over its efficacy in elderly populations still does exist.

Decubitus ulcers can be prevented in a variety of ways (24) and this is far to be preferred to attempts at treatment of infected ulcers. In our Wisconsin data, during the initial prevalence study, most of the decubitus ulcers were in residents of one nursing home. This was the home with the least well developed infection control program initially. It was our impression that attitudes of the nursing staff, especially the chief of the nursing service, were crucial to the success of the infection control program. The most positive attitudes about carrying out a comprehensive infection control program were associated with the resident populations with no decubiti and infrequent use of indwelling catheters.

Major problems can occur with cross infection or outbreak situations. Of these, clusters of gastroenteritis can be the most important (25), spread of antibiotic resistant organisms and introduction from nursing home to acute care hospitals can be the most problematic, influenza and other respiratory pathogens - even tuberculosis - can be the most risk for the staff (26,27,28) and clusters of conjunctivitis are probably the most preventable using good handwashing technique. Standard food and water sanitation, hygiene practices and prompt recognition and appropriate isolation of residents with gastroenteritis is probably the best prevention for this problem. Good communication between acute care hospitals and nursing homes, coupled with appropriate microbiological techniques and a variety of containment isolation strategies, is probably the best that can be done with resistant organisms. Avoidance of inappropriate antibiotic use in nursing home residents can also minimize the emergence of resistant strains (17,18,29,30,31,32). The issue pneumonia prevention has been discussed along with the use of influenza vaccination. Several (23,28) have stressed the importance of good employee health policy regarding use of influenza vaccine, avoidance of work when ill with respiratory infections and attention to tuberculin status. Stead and his colleagues in Arkansas (27) have demonstrated the importance of the TB skin test status and change in status among nursing home residents in that state and have encouraged early recognition and prevention INH therapy even in elderly residents.

CONCLUSION

The prevalence surveys and the few ongoing surveillance reports from selected nursing homes around the country are beginning to expand our understanding of infections among residents in nursing homes. This resident population will continue to grow substantially throughout the rest of this century and well into the next. Therefore, it is important for us to build on our current knowledge and apply what currently seems rational. Strategies that prevent cross infection, minimize spread of community outbreaks of infection with judicious use of immunization and isolation programs, and utilize what is known about the epidemiology and natural history of specific infections among nursing home residents to prevent acquisition of infections need to be implemented in all nursing homes.

The elements of a nursing home infection and control strategy that make the most current sense are as follows:

1. Regular surveillance to define and monitor the problem of infections.

2. A small infection control group consisting of the nursing director, one interested physician and the infection control nurse with authority to monitor infections and policy, and intervene as necessary.

3. An immunization program:

   - Annual influenza immunization
   - One-time pneumococcal immunization
   - Adequate knowledge to TB status

4. Appropriate strategies and policies for:

   - Inservice and continuing education of employees
   - Adequate staffing levels
   - Appropriate isolation policies - adapt CDC guidelines
   - Appropriate skin care, toileting and catheter guidelines
   - An Employee Health policy that allows for absenteeism
   - Adequate handwashing facilities and emphasis for residents and employees
   - Avoidance of patient clustering as needed to prevent spread of contagious infections

The Centers for Disease Control should consider developing - perhaps in cooperation with the National Institutes on Aging - a cooperative national surveillance network of nursing homes across the country. This project could be developed on a voluntary basis and patterned after the National Nosocomial Infections Study (NNIS) in acute care hospitals. Standard definitions of and categories for infections could be developed. Continued separation of the infections among residents in nursing homes into nursing home associated, community acquired and nosocomial (for hospital associated infection only) should help develop our knowledge in this area more scientifically and continue to highlight the intrinsic differences between nursing home resident populations and acute care hospital patient populations.

LITERATURED CITED

1. Metropolitan Insurance Companies, United States population outlook, Statistical Bull., 65(1):16-19 (1984).
2. R.W. Haley, D.H. Culver, J.W. White, et al., The efficacy of surveillance and control programs in preventing nosocomial infections in U.S. hospitals, Am. J. Epidemiol. 121:182-205 (1985).
3. L.E. Price, F.A. Sarubbi, and W.A. Rutala, Infection control programs in 12 North Carolina extended care facilities, Infect. Control 6:437-441 (1985).
4. K.B. Crossley, P. Irvine, D.J. Kaszar, et al., Infection control practices in Minnesota nursing homes, J.A.M.A. 254:2918-2921 (1985).

5. R.F. Khabbaz, and J.H. Tenney, Infection control in Maryland nursing homes, Infect. Control Hosp. Epidemiol. 9(4):159-162 (1988).
6. W.E. Scheckler, and P.J. Peterson, Infections and infection control among residents of eight rural Wisconsin nursing homes, Arch. Intern. Med. 146:1981-1984 (1986).
7. D. Vlahov, J.H. Tenney, K.W. Cervino, and D.K. Shamer, Routine surveillance for infections in nursing homes: experience at two facilities, Am. J. Infect. Control 15(2):47-53 (1987).
8. J.S. Powers, F.T. Billings, D. Behrendt, and M.C. Burger, Antecedent factors in urinary tract infections among nursing home patients, South. Med. J. 81(6):734-735 (1988).
9. L.E. Nicolle, W.J. Mayhew, and L. Bryan, Prospective randomized comparison of therapy and no therapy for asymptomatic bacteriuria in institutionalized elderly women, Am. J. Med. 83:27-33 (1987).
10. J.G. Ouslander, B. Greengold, and S. Chen, External catheter use and urinary tract infections among incontinent male nursing home patients, J. Am. Geriatr. Soc. 35(12):1063-1070 (1987).
11. T.J. Marrie, H. Durant, and C. Kwan, Nursing home-acquired pneumonia. A case-control study, J. Am. Geriatr. Soc. 34:697-702 (1986).
12. G.A. Roselle, Nosocomial and nursing home-acquired pneumonia. Recent therapeutic advances, Postgrad. Med. 81(1):131-136 (1987).
13. N.H. Arden, P.A. Patriarca, M.B. Fasano, et al., The roles of vaccination and amantadine prophylaxis in controlling an outbreak of influenza A (H3N2) in a nursing home, Arch. Intern. Med. 148:865-868 (1988).
14. P.A. Gross, M. Rodstein, J.R. LaMontagne, et al., Epidemiology of acute respiratory illness during an influenza outbreak in a nursing home. A prospective study, Ann. Intern. Med. 148:559-561 (1988).
15. P.A. Patriarca, N.H. Arden, J.P. Koplan, and R.A. Goodman, Prevention and control of type A influenza infections in nursing homes. Benefits and costs of four approaches using vaccination and amantadine, Ann. Intern. Med. 107:732-740 (1987).
16. P.A. Patriarca, J.A. Weber, R.A. Parker, et al., Risk factors for outbreaks of influenza in nursing homes. A case-control study, Am. J. Epidemiol. 124(1):114-119 (1986).
17. J.G. Zimmer, D.W. Bentley, W.M. Valenti, and N.M. Watson, Systemic antibiotic use in nursing homes. A quality assessment, J. Am. Geriatr. Soc. 34:703-710 (1986).
18. S.R. Jones, D.F. Parker, E.S. Liebow, et al., Appropriateness of antibiotic therapy in long-term care facilities, Am. J. Med. 83:499-502 (1987).
19. R.A. Garibaldi, S. Brodine, and S. Matsumiya, Infections among patients in nursing homes. Policies, prevalence, and problems, N. Engl. J. Med. 305:731-735 (1981).
20. J.A. Boscia, W.D. Kobasa, R.A. Knight, et al., Epidemiology of bacteriuria in an elderly ambulatory population, Am. J. Med. 80:208-214 (1986).
21. C.M. Kunin, Q.F. Chin, and S. Chambers, Morbidity and mortality associated with indwelling urinary catheters in elderly patients in a nursing home - confounding due to the presence of associated diseases, J. Am. Geriatr. Soc. 35:1001-1006 (1987).

22. P.A. Gross, G.V. Quinnan, M. Rodstein, et al., Association of influenza immunization with reduction in mortality in an elderly population. A prospective study, <u>Arch</u>. <u>Intern</u>. <u>Med</u>. 148:562-565 (1988).
23. Update: Pneumococcal Polysaccharide Vaccine Usage - United States, <u>M.M.W.R</u>. 33(20):273-281 (1984).
24. B.L. Moody, J.E. Fanale, M. Thompson, et al., Impact of staff education on pressure sore development in elderly hospitalized patients, <u>Arch</u>. <u>Intern</u>. <u>Med</u>. 148:2241-2243 (1988).
25. C.A. Ryan, R.V. Tauxe, G.W. Hosek, et al., <u>Escherichia</u> <u>coli</u> 0157:H7 diarrhea in a nursing home: clinical, epidemiological, and pathological findings, <u>J</u>. <u>Inf</u>. <u>Dis</u>. 154(4):631-638 (1986).
26. M.C. Creditor, E.C. Smith, J.B. Gallai, et al., Tuberculosis, tuberculin reactivity, and delayed cutaneous hypersensitivity in nursing home residents, <u>J</u>. <u>Gerontol</u>. 43(4):M97-M100 (1988).
27. W.W. Stead, and T. To, The significance of the tuberculin skin test in elderly persons, <u>Ann</u>. <u>Intern</u>. <u>Med</u>. 107:837-842 (1987).
28. P.W. Smith, Consensus conference on nosocomial infections in long-term care facilities, <u>Am</u>. <u>J</u>. <u>Infect</u>. <u>Control</u> 15(3):97-100 (1987).
29. M.J. Zervos, M.S. Terpenning, D.R. Schaberg, et al., High-level aminoglycoside-resistant enterococci. Colonization of nursing home and acute care hospital patients, <u>Arch</u>. <u>Intern</u>. <u>Med</u>. 147:1591-1594 (1987).
30. D.M. Shlaes, M.H. Lehman, C.A. Currie-McCumber, et al., Prevalence of colonization with antibiotic resistant gram-negative bacilli in a nursing home care unit: the importance of cross-colonization as documented by plasmid analysis, <u>Infect</u>. <u>Control</u> 7(11):538-545 (1986).
31. G.A. Storch, J.L. Radcliff, P.L. Meyer, and J.H. Hinrichs, Methicillin-resistant <u>Staphylococcus</u> <u>aureus</u> in a nursing home, <u>Infect</u>. <u>Control</u> 8(1):24-29 (1987).
32. R.A. Weinstein, Resistant bacteria and infection control in the nursing home and hospital, <u>Bull</u>. <u>N.Y</u>. <u>Acad</u>. <u>Med</u>. 63(3):337-344 (1987).

STERILIZATION AND DISINFECTION STRATEGIES USED IN HOSPITALS

Martin S. Favero

Centers for Disease Control

Atlanta, Georgia 30333

The purpose of this paper is to discuss the science and art of sterilization and disinfection practices in health care facilities in the United States. In addition, the most current procedures that are used to test germicidal efficacy will be discussed.

Currently, health-care facilities in the United States perform sterilization by heat, such as steam under pressure or dry heat, ethlyne oxide gas, or liquid chemical germicides. The emphasis of this paper will be sterilization or disinfection accomplished by liquid chemical germicides.

In the United States, although there are a number of commercially available liquid chemical germicides that can be used for sterilization, this practice is infrequently performed. Rather, these liquid chemical formulations are used primarily for disinfection.

Definitions

The definition of "sterilization" changes depending on the vantage point from which it is viewed. I choose to view this term somewhat like a hologram and will define it in the context of the state of sterilization; the procedure of sterilization; and the application of sterilization.

A device is considered to be sterile when it is completely free of all living microorganisms. This state of sterility is the objective of the sterilization procedure and when viewed in this context the definition, for all practical purposes, is an absolute one. An item is either sterile or it is not.

A sterilization procedure is one that kills all microorganisms including high numbers of bacterial spores and can be accomplished by heat, ethylene oxide gas, radiation and certain chemical germicides - notably those approved by the U.S. Environmental Protection Agency (EPA) as sterilant/sporicides. Sterilization procedures from an operational standpoint, cannot be categorically defined. Rather the procedure is defined as a process, after which, the probability of a microorganism surviving on an item subjected to sterilization is less than one chance in a million (10^{-6}). It is this approach that is used by the medical device industry when large quantities of medical devices are sterilized. Some of the following criteria are used in the production and labeling of a sterile device:

- o Good manufacturing practices
- o Validated sterilization process
- o Use of biological indicators
- o Sterility testing of a subsample of the batch subjected to the sterilization procedure
- o Process controls
- o Quality control on materials
- o Post-sterilization testing of devices for function

The _application_ of a sterilization process takes into account additional considerations. This approach involves the strategy associated with a particular medical device or solution in the context of its exposure to humans. I will point out later the classification used by Dr. Earl Spalding for devices that are exposed to sterile areas of the body such as blood that require sterilization; while devices that touch mucous membranes may either be sterilized or disinfected; and those devices or items that only touch skin that can be sanitized or cleaned with soap and water.

The bacterial spore is the type of microorganism used as a biological challenge and is a major component among several tests used to judge the efficacy of a sterilization process. Spores are also used as a biological monitor of sterilization processes in health-care facilities. Liquid chemical germicides used for sterilization in health-care facilities cannot be and are not monitored with biological indicators.

However, standard tests for determining sporicidal capability of sterilization procedures in general and liquid chemical germicides in particular do make use of standard well-described suspensions of bacterial spores. Standardization is necessary in order to compare the germicidal activity between formulations of chemical germicides as well as establishing minimum criteria for being able to be qualified as a sporicide. The one notable disadvantage of this approach is that laboratory grown spore suspensions bear little resemblance in their resistance capabilities to naturally occurring spore populations which are representative of the extreme microbiologic burden to which the sterilization procedures are targeted. This latter degree of resistance can only be simulated and an extreme degree of resistance may have to be sacrificed for a subcultured, standardized spore suspension so that accuracy and precision of test results can be high.

Factors that influence inactivation of microorganisms

There are a number of factors that influence inactivation of microorganisms by liquid chemical germicides that must be controlled and specified in efficacy tests. Among these factors are:

- o The type of microorganisms - bacterial spores are more resistant than mycobacteria, fungi, vegetative bacteria, and viruses; some types of viruses are more resistant to certain germicides than others.

o   The number of microorganisms - all other factors being equal, the higher the number of microorganisms of a certain type present on a device, the longer it takes to inactivate this microbial population.

o   The intrinsic resistance of microorganisms - naturally occurring microorganisms including gram negative water bacteria such as <u>Pseudomonas</u> <u>aeruginosa</u>, non-tuberculous mycobacteria, and bacterial spores are significantly more resistant in their natural occurring state when exposed to physical and chemical agents than are the same microorganisms once they are subcultured. In efforts to standardize efficacy tests there are significant attempts to standardize the cultures used and invariably these cultures are ones that are perpetuated and retained in laboratories.

o   Other factors - temperature, organic load, pH, germicide concentration, and inorganic load are additional factors that influence the rate of inactivation of microorganisms by chemical germicides. With some chemicals, such as iodophors, the amount of water present may be extremely important; more concentrated solutions of iodophors may contain less free iodine than those that are more dilute. Factors such as this must be considered both in efficacy testing as well as use.

Efficacy Testing

There are basically three approaches to testing efficacy of chemical germicides that are used or can be used as sporicide sterilants. These are:

o   AOAC Tests

o   Survival time - rate

o   Artificially contaminated devices

o   Naturally contaminated devices

AOAC tests are very standardized, make use of laboratory-grown microbial cultures and are performed under strict laboratory conditions. The results of these qualitative tests are expressed as positive or negative. Likewise, there are other procedures for determining the inactivation rate of microorganisms when exposed to chemical germicides. These tests often make use of quantitative data in the form of survival curves and also employ laboratory cultures grown under controlled conditions.

There are many studies in the scientific literature that describe particular medical devices that are artificially contaminated with a variety of laboratory cultured microorganisms. The devices are then subjected to a sterilization or disinfection procedure after which appropriate microbiologic assays are performed to determine the efficacy of the procedure. This approach has the advantage of the disinfection procedures approaching the actual use conditions in the health care facility but suffers oftentimes from lack of precision and accuracy.

Finally, there are those tests that involve in-use medical devices subjected to chemical sterilization or disinfection in health care facilities and subsequently are assayed for effectiveness of the procedures. In this instance the microbial load on the device will be the natural microbial populations encountered during use on patients in a health care facility. The advantage of this approach is that it reflects natural situations actually encountered during use of the device. However, the inherent lack of precision and accuracy due to varying types of microorganisms as well as the absence of a worst case challenge, oftentimes relegates this type of test to one that is done only in special circumstances in health-care facilities but not one that is performed to test a chemical germicide for efficacy.

Dr. Earl Spaulding proposed almost two decades ago a method of classifying risk levels of medical devices and surfaces in health care facilities and concomitantly categorized the types of chemical germicides that should be used in strategies for sterilizing or disinfecting each particular category of device or surface. This system differs from the one used by the Environmental Protection Agency but is one that has been used by a number of individuals who have written chapters on disinfection and sterilization and is a system that I and my colleagues at the Centers for Disease Control have used for a number of years in recommending guidelines for environmental control especially as they relate to disinfection and sterilization strategies.

Briefly, surfaces or medical devices can be categorized as:

o   Critical surfaces - penetrate skin or mucous membranes - sterilization is required.

o   Semi-critical surfaces - touch mucuous membranes; sterilization or high-level disinfection can be used.

o   Non-critical surfaces - these surfaces or devices touch, but do not penetrate, intact skin; strategies can range from low to intermediate level disinfection or in many cases cleaning with soap and water.

Dr. Spaulding classified chemical germicides by activity level in the following manner:

o   High-level disinfection - a procedure that kills all vegetative microorganisms but not high numbers of bacterial spores: these chemical germicides quite often are those that are registered with the EPA as sterilant/sporicides. An integral part of the definition of this type of chemical germicide is that it is one that is capable of sterilization if the contact time is long enough, e.g., that it is sporocidal at use dilution; if contact time is relatively short then the chemical germicide process is referred to as high-level disinfection.

o   Intermediate-level disinfection - a procedure that kills vegetative bacteria including M. tuberculosis, all fungi and most viruses. These chemical germicides quite often correspond to EPA-approved "hospital disinfectants" that are also "tuberculocidal".

o   Low-level disinfectants - a procedure that kills most vegetative bacteria except M. tuberculosis, some fungi and some viruses. These chemical germicides, similar to the ones approved by EPA as "hospital disinfectants" or "sanitizers".

This system for classifying devices and strategies for disinfection and sterilization is quite conservative. There is a direct relationship between the degree of conservatism as expressed by the probability of a microorganism surviving a particular procedure and microbicidal activity. From a theoretical standpoint the likelihood of infection resulting from the use of a device after a treatment is shown in Table I. Sterilization accomplished by either steam autoclaving or dry heat has a probability of one in a million of a survivor being present after the procedure if the procedure had initially been challenged with 10^6 highly resistant bacterial spores. The probability of infection being transmitted from an item that was subjected to this type of procedure, assuming it had been carried out properly, is zero. Correspondingly, the probability of contamination and the estimated theoretical probability of infection associated with procedures of chemical sterilization, high-level disinfection and low-level disinfection, increase as the germicidal activity of the procedure decreases.

My major point here is that a process of chemical sterilization at best would have one thousand times or three orders of magnitude less reliability than conventional heat sterilization. From a practical standpoint this means that there is less insurance in such a procedure and if and when mistakes are made there is a more likely chance of failure than there is with heat sterilization.

TABLE I
Theoretical Considerations

Procedure	Probability of Contamination	Probability of Infection
Heat Sterilization	10^{-6}	0
Chemical Sterilization	10^{-3}	0 to ?
High-level Disinfection	1 to 10^{-2}	1 to ?
Low-level Disinfection	1	1 to ?

In the last twenty years there has been a gradual shift by health care facilities from using a strategy of extreme conservatism to policies and procedures that appear to be safe and effective but are not as conservative. For example, there are a variety of medical devices which twenty years ago would have been categorically recommended to be sterilized in between patient use. Today, many of those devices are actually submitted to a procedure of high-level disinfection rather than sterilization without any apparent problems associated with this practice except when errors are made in the procedure. As I mentioned previously it appears that errors, when they are made, have much more of a consequence in a procedure of high-level disinfection than a sterilization procedure. Lensed instruments such as endoscopes which come in contact with mucous membranes and others that actually penetrate skin and mucous membranes can be either sterilized or disinfected. It appears that most practitioners in most health care facilities have elected to submit these devices to a cleaning procedure followed by a procedure of high-level disinfection. Very few of these health care facilities practice uniformly the procedure of sterilization of these devices in between patient use. It also appears that if this procedure is done according to a good protocol that there are no significant problems of infection transmission. However, it is also evident that when mistakes are made during these procedures infections can be transmitted from patient to patient.

Another example of the changing attitude of conservatism is the artificial kidney - the disposable hemodialyzer. This device is manufactured and labled sterile; it is disposable and since the mid-1960s nephrologists have been reprocessing these devices for use on the same patient. It is evident that if the protocols of reprocessing are performed adequately that there are no problems of infections by dialyzing patients. The practice of reusing dialyzers has increased from about 18% of centers in the U.S. in 1976 to 64% of centers in 1987. The reprocessing procedures primarily make use of aqueous solutions of formaldehyde ranging from 2-4%. Contact time at room temperature is approximately 48 hours. By no stretch of the imagination could one describe this regime as a sterilization procedure. At best it is a high-level disinfection procedure and in some cases where 2% formaldehyde is used at 20°C the procedure may only be intermediate level disinfection. The primary force behind this change in attitude is one of cost. It has been shown that centers that are privately owned and operated for profit are much more likely to have a dialyzer reprocessing program than centers that are operated for nonprofit and are associated with hospitals.

Future Needs

There are a number of research needs and policy changes that are needed for the future. The American Society for Microbiology convened a symposium on the efficacy of chemical germicides used in the health care field. This symposium identified those sporicidal tests which are used by regulatory agencies and in the scientific community that require updating with respect to accuracy and precision, and those that have the capability of predicting the efficacy of chemical germicides in a standardized fashion. From the standpoint of the health-care professional a simplified method is needed to allow them to make judgments on correct use of chemical germicides that are in the market place. Currently there are approximately 300 active ingredients that have been registered with the EPA and approved for use in chemical germicides. About 14 of them are in 92% of the registered products and there are approximately 14,000 products. This situation has oftentimes stimulated a great deal of exaggeration in the marketing of germicides to health-care professionals and in some cases outright false or misleading advertising. Although this element does not deal directly with the testing of the efficacy of chemical germicides it plays a significant role in the realistic use of chemical germicides in these facilities.

Bibliography

1.   Disinfection, Sterilization and Preservation, 1983.   3rd Edition, Seymour S. Block, Ed., Lea & Febiger, Philadelphia.
2.   Favero, MS. Chemical disinfection of medical and surgical materials. In: Disinfection, Sterilization and Preservation, 3rd Edition, Seymour S. Block, Ed., pp. 469-492, Lea & Febiger, Philadelphia, 1983.
3.   Favero, MS. Sterilization, disinfection, and antisepsis in the hospital. From: Manual of Clinical Microbiology, 4th Ed., Chapter 13, pp. 129-137, American Society for Microbiology, 1985.

HOW SAFE IS THE BLOOD SUPPLY IN THE UNITED STATES?

Roger Y. Dodd

American Red Cross
Jerome H. Holland Laboratory
15601 Crabbs Branch Way
Rockville MD 20855

I. INTRODUCTION

It has long been known that transfusion of blood or its derivatives may also result in the inadvertent transmission of infectious agents, particularly those associated with a silent carrier state. For many years, hepatitis and syphilis were the only diseases which were recognized as transmissible by this route, but there appeared to be relatively little anxiety among the general public. The recognition of AIDS among recipients of clotting factor concentrates and single donor transfusions was, in contrast, widely publicised and led to considerable and continuing concern. Indeed, the association between blood transfusion and AIDS became so firm that a significant proportion of the population mistakenly believed that AIDS could be contracted as a result of giving blood. The need to reduce transfusion associated AIDS led to a new and more intensive focus on the whole issue of the safety of the blood supply. In fact, in the three years following the introduction of HIV antibody screening, three additional tests were implemented by blood collection agencies as measures to reduce the transmission of other viruses. In addition, there have been demonstrable improvements in the methods used to screen the donor population and there is real hope that inactivation of viral or microbial contaminants in cellular blood products will be achieved. It will be shown that the chance of being infected by blood transfusion has never been lower than it is now. Finally, physicians have recognized the need to review the risks and benefits of every transfusion and to seek safe alternatives wherever possible.

II. TRANSFUSION TRANSMITTED INFECTION

A. Single Donor Products

In the United States, approximately thirteen million donations of blood are made each year. These donations are exclusively from volunteers; paid donations have been eliminated. These donations are provided by about eight million individuals, many of whom give more than once per year. The donations are used to treat about four million patients, representing a wide variety of medical needs. Because these needs reflect differing requirements for blood components, almost all blood donations are used to prepare several components, including packed red cells, platelet concentrates, fresh frozen plasma and cryoprecipitate (a source of clotting factors). There are also a limited number of treatments with products prepared by apheresis; primarily platelets and occasionally granulocytes.

The majority of transfusions of single donor products are given to patients who are 65 or more years of age. There is also a significant amount of transfusion in the very early years, particularly for infants of very low birthweight. Although modern surgical approaches are minimizing the need for transfusion, other therapies are resulting in an increased demand for selected products. For example, chemotherapy for the treatment of cancer can compromise marrow function and generate a need for transfusion, especially of platelets. Additionally, transplantation, particularly of the liver or bone marrow, generates a need for large volumes of transfusion products.

B. Pooled Plasma Products

One of the major advances in the treatment of hemophilia has been the development of concentrates of clotting factors derived from pooled plasma. In addition, other therapeutic or prophylactic products, such as albumin and immune serum globulins are prepared from this source. The plasma which is used for fractionation is derived from two sources. The majority comes from paid donors who are plasmapheresed and give about 400 mL of plasma at each donation, which may be as frequent as twice per week. Approximately 8 million such donations are given each year in the United States. Plasma may also be recovered from routine donations. In order to process the plasma into components, large pools are made; these pools may contain plasma from as many as 20,000 different donors. Thus, even a relatively small prevalence of infection among the donors can lead to significant risk of contamination of the pool. Plasma is fractionated using a modified cold ethanol procedure with a number of additional steps for particular components. As discussed below, plasma products are routinely treated to eliminate or reduce any residual infectivity.

Clotting factor concentrates are mainly used to treat patients with hemophilia A, who require Factor VIII, or the less frequent hemophilia B, who require Factor IX. In total, it is estimated that there are 16 to 20 thousand hemophilia patients in the United States. Albumin is routinely used as a volume expander. Immune serum globulin preparations are used for the support of immunodeficient individuals and for specific prophylaxis against a number of infections. There are a number of additional uses for special preparations of immunoglobulins designed for intravenous administration.

C. Infectious Agents Transmitted by Transfusion.

Blood is collected from carefully screened, healthy donors who must provide an acceptable medical history and who must lack symptoms of disease or infection at the time of donation. Thus, although many infectious agents may be found in the blood, the majority are not transmitted by transfusion because they are only present in the blood after disease symptoms or fever have appeared. In particular, this means that transmission of bacterial infection from donor to recipient is extremely rare, since bacteremia usually reflects serious acute infection. Although it might be anticipated that intracellular bacterial pathogens, such as those responsible for brucellosis and tularemia, might be transmissible by this route, this does not appear to be a problem in practice. Treponema pallidum is transmissible by blood, but current methods used to collect and store blood have virtually eliminated this risk; even so, serologic tests for syphilis continue to be required. There are infrequent cases of infection with other bacteria, apparently transmitted from infected donors or as a result of contamination with skin bacteria which may enter the collection bag at the time of phlebotomy.

There is little doubt that viruses are the major potential contaminant of blood, at least in the United States. Hepatitis B virus (HBV) and the as yet uncharacterised agent or agents of non-A, non-B (NANB) hepatitis have been the prototypes of transfusion-transmitted viruses. In terms of frequency of infection, NANB hepatitis is the most important of the diseases to affect blood recipients. However, the severity of AIDS and its emotional connotations have resulted in most attention being focused upon the human immunodeficiency virus (HIV). In common with other retroviruses, infection leads to integration of the viral genome into that of the host and infection is lifelong. Productive infection also seems to be persistent. Other human retroviruses have also been described (14,63) and it is clear that HTLV-I can be transmitted by blood transfusion (51,54,55) and the frequency of the closely related HTLV-II among drug addicts strongly suggest that it is also transmissible via the blood. It is, of course reasonable to expect that additional human retroviruses will be recognized and that they will prove to be transmissible via blood transfusion.

Herpesviruses also cause persistent infection and at least three, cytomegalovirus, Epstein-Barr virus and the recently described human herpes virus 6 (2,65), are able to infect leukocytes. However, CMV is the only herpes virus which is known to result in disease as a consequence of transmission by transfusion. In an extensive review, Henle (39) was only able to find a single case in which EBV was thought to have been the cause of a transfusion associated disease. It is probably too early to define the possible role of HHV 6 in this context.

Parasites are also transmissible by transfusion. Most notably, the various forms of malaria are a major problem in many parts of the world. In the United States, however, transfusion malaria does not seem to be a problem, with fewer than ten cases reported annually. This is largely attributable to careful screening of donors to assure that they have not resided in, or visited, areas where malaria is endemic. Interestingly, there are a few reports of transfusion associated disease with the closely related babesia parasite in the Northeast United States (59). Chagas' disease is a serious and frequent transfusion associated infection in South America and there is some concern that certain immigrant populations, particularly in the Southwest, may be at risk of introducing infection into the U.S. blood supply.

D. Approaches to Blood Safety

Currently, the safety of the supply of single donor products is dependent upon two key lines of defense, donor selection and laboratory testing. The former probably makes the greatest overall contribution to the reduction of transfusion transmitted infection. The efficacy of donor selection is perhaps not fully appreciated, but it can be credited for reducing the incidence of clinically recognized posttransfusion hepatitis at least eight fold, even prior to the introduction of testing (30). In addition, the proportion of newly presenting blood donors who are found to be infected with HIV is less than 10 percent of the proportion which would be expected in a random sampling of the U.S. population. There is clear evidence that continuing improvement in donor selection and medical screening is leading to a concomitant increase in the overall safety of the donor population. Selection is based upon two levels of activity. First, certain groups of donors have clearly been found to offer increased risk of infection and collection of blood from these groups is stopped. The most outstanding examples of the success of this approach have been the exclusion of collection from prisoners and from paid donors.

The second level of selection is directed towards individuals. Presenting donors are given extensive educational materials which stress the dangers of disease transmission by transfusion and which define characteristics or behaviors which make a person ineligible to donate. These materials have also been broadly publicised. The intent is that persons with risk factors for HIV, HBV, or other forms of infection should defer themselves - i.e. should decline to give blood. This option is backed up by the confidential unit exclusion mechanism, which requires that a donor specify, in a confidential fashion, that his or her blood donation should, or should not be used for treatment of patients. Presenting donors have a brief medical examination and must complete a medical history interview. The history seeks specific risk factors for infection, current symptoms and previous disease history. The donor is asked to affirm, by signature, that the answers are truthful and that he or she has no risk for HIV infection.

Laboratory testing identifies the majority of those few infectious blood donations which escape the donor screening process. There are two distinct types of laboratory testing. Specific tests are designed to identify the footprints of a given virus in the donor; HBsAg for HBV and antibodies to HIV, for example. Surrogate testing is also used where a specific viral marker is not available, as in the case of NANB; the test identifies a population of individuals which has been shown to offer increased risk of transmitting the infection in question. Testing will be described in context below.

E. Approaches to Plasma Product Safety

A third level of defense is the antiviral treatments which are routinely applied to therapeutic products prepared from pooled plasma (34). More specifically, preparations of Factor VIII and Factor IX are subjected to one or more of a number of procedures.

Heating, usually, but not always in the dried state has proven quite efficacious in reduction or elimination of HIV infectivity, particularly when more recent protocols with increased temperatures and times have been used. The relatively recent introduction of treatment with tributyl phosphate and cholate has also been extremely effective in eliminating the infectivity of enveloped viruses, including HIV. Finally, specific purification protocols using monoclonal antibodies to select for clotting factors and reject other contaminants has also contributed to safety. Current data suggest that plasma products are extremely safe. Mannucci has reported on a total of almost 150 uninfected, previously untreated patients in 9 separate studies, none of whom developed HIV infection as a result of treatment with current, virally inactivated products (50). However, it is still unclear whether these treatments have been fully successful in eliminating the transmission of NANB hepatitis. There is no evidence that other therapeutic products prepared from pooled plasma are capable of transmitting HIV. More specifically, albumin is completely safe because of the combination of high levels of ethanol used during its preparation and the routine pasteurization of this product, which is heated, in the liquid state, for 10 hours at 60°C. Immune serum globulin preparations, although not heated are prepared using a high concentration of ethanol and are noninfectious. There has been some controversy about the potential infectivity of immune globulins prepared for intravenous use, but these products are also virus free and have not been implicated in the transmission of HIV infection. It should be noted that there have been reports of passive transmission of HIV antibody to recipients of intravenous immune globulins and specific hepatitis B immune globulin. This could clearly be differentiated from a true infection with HIV by observing the decay of the passive antibody level as the immunoglobulins are lost from the circulation.

F. Physician's Responsibility for use of Blood.

The purpose of blood transfusion is to replace needed functions which have been lost as a result of hemorrhage, loss of hematopoietic capacity or dysfunction of selected blood components; in and of itself, blood is not appropriate for management of fluid loss. Most frequently, surgical and traumatic blood loss result in the need to replace oxygen carrying capacity in the patient. This function is best achieved by transfusion of packed red cells. In the past, a particular circulating hemoglobin level, most often 10 g/dL, was selected as the trigger for a transfusion, irrespective of the physiologic status of the patient. It can be argued that this level is arbitrary and may have led to inappropriate transfusions in many cases. The use of such a rigid transfusion trigger should be replaced by a more conservative approach in which the physiologic needs of the patient are evaluated in order to establish the need for transfusion. It is also important to consider what a reasonable therapeutic dose of red cells should be. In general, a single unit of blood does not provide an adequate increment of oxygen carrying capacity and thus offers risk of infection, but little benefit. This is not to say that a minimum of two units should be given, but rather to suggest that a patient who is thought to need only a single blood unit may not need to be transfused at all.

A recent NIH Consensus Development Conference reviewed the indications for transfusion of fresh frozen plasma and concluded that there were very few circumstances in which this product is indicated at all. Nevertheless, it continues to be used, but physicians should certainly consider the use of other therapies before prescribing fresh frozen plasma. Perhaps most pernicious is the practice of reconstituting whole blood by transfusing packed red cells and plasma. Since the plasma and red cells will most likely come from different donors, the risk of infection is doubled relative to a unit of packed cells.

Platelets are most frequently transfused to patients with bleeding disorders as a result of chemotherapy or other cytotoxic procedures which have affected marrow function. A therapeutic dose of platelets is obtained from 6 to 7 standard units, each of which is prepared from a single blood donation. This is also equivalent to a unit of single donor platelets, prepared by apheresis from one donor. The latter product would clearly offer a lower risk of infection, but is not uniformly available.

In the past, hemophilia patients requiring therapy with clotting factor concentrates have been at high risk of infection with HIV. Currently, however, viral inactivation protocols have resulted in great increases in the safety of these products. The National

Hemophilia Foundation has stated that "The risks of withholding factor treatment far outweigh the risks of treatment, but health-care providers should educate patients to use appropriate doses of clotting factor to minimize overuse and to contain costs." It was also recommended that Desmopressin (DDAVP, 1-desamino-8-D- arginine vasopressin) should be used wherever possible by patients with mild or moderate hemophilia A. An alternative to concentrates may be the use of cryoprecipitate prepared from one, or a small number of well-screened and repeatedly tested donors. However, it is recognized that factor concentrates which have been heat treated, particularly by the most recent methods, and those which have been solvent-detergent treated and/or purified using monoclonal antibodies are at substantially reduced risk of transmitting HIV (24).

While it is clearly important to evaluate the need for transfusion of homologous blood, it is necessary to consider alternatives to traditional transfusion wherever possible. In particular, autologous transfusion should be used in all appropriate circumstances. Three basic approaches are available. Predeposit autologous transfusion may be used for elective surgery. The prospective patient is bled several times during the four weeks or so prior to surgery and his or her blood is stored for use as needed during or after the operative procedure. It is permissible to bleed the patient more frequently than the once every 8 weeks which is sanctioned for routine blood donors; in addition, blood may be drawn in patients with a hematocrit of 34% or greater. Two to three units of blood can usually be collected without compromising the patient. It may be possible to reduce the need for transfusion or to increase the availability of autologous blood by using recombinant erythropoietin; a drug which is currently in clinical trials for this purpose. One controversial aspect of predeposit autologous transfusion is the management of blood which is not needed by the patient. Should the blood be returned to the blood center or transfusion service for use in other patients, or should it be discarded? There is no clear policy in this area (8). However, in order to be transfused to others, the blood unit must meet all standards appropriate for homologous blood units. Even so, some feel that there may be an increased risk of infectivity since the person who deposits blood for his or her own use is logically more like a hospitalized patient and a first time donor. Both of these groups tend to have higher frequencies of infectious disease markers than do routine blood donors.

The second form of autologous transfusion is hemodilution, in which blood is drawn from the patient at the initiation of surgery and the volume is replaced with saline. The blood is transfused back to the patient on an as- needed basis. This procedure has the benefits of improving the hemodynamic status of the patient and of reducing the loss of cellular components in the event of intraoperative blood loss.

Finally, intraoperative salvage of blood is of increasing importance where circumstances permit. Blood is recovered from the operative site and the red cells are recovered, washed and reinfused. This procedure can markedly reduce or eliminate the need for homologous transfusion. However, intraoperative salvage cannot be used in abdominal surgery or in other situations where there is a risk of contamination of the operative field. Similarly, there is a risk of promoting metastasis in oncologic surgery.

III. BACTERIA

A. Treponema pallidum

Transmission of syphilis by blood transfusion was first reported in 1915. In the early days of blood transfusion, many such cases were reported. The potential for transmission is essentially confined to the early stages of infection. However, the development of methods for the collection of blood into anticoagulants, followed by storage at refrigerator temperatures effectively eliminated the risk of transmission, since the spirochete is inactivated within a day or two. Even in 1965, a review of the literature revealed no cases of transfusion transmitted syphilis attributable to blood which had been properly collected and stored (75). Since platelets are routinely stored at room temperature, there may be some potential for transmission, but this does not seem to have been realized. At the same time, it must be recognized that every blood and plasma donation is subjected to a serologic test for syphilis. Such tests are almost always cardiolipin based tests for reagins. Reactivity rates in donor populations are generally less than 0.1% and a significant proportion of these (perhaps 50%) are false positives, as defined by confirmation using specific treponemal tests (70). Clearly, however, serologic tests for syphilis cannot provide any protection prior to the appearance of antibodies, when there is most risk of infectivity.

Since blood collection and storage technology essentially eliminate the risk of transfusion associated treponemal infection, and since current tests are probably ineffective, it is reasonable to ask why blood collections continue to be tested. At this time, federal regulations require this testing. There has been consideration of eliminating this requirement, consideration which was interrupted by concern about the AIDS epidemic and the concept that syphilis testing might perhaps identify individuals at risk for AIDS. Interestingly, the standards for blood banking which are developed and published by the American Association of Blood Banks do not require the performance of a serological test for syphilis but Federal regulations take precedence. In summary, however, with or without serologic testing, the chance of being infected with T. pallidum as a result of blood transfusion is vanishingly small.

B. Yersinia enterocolitica

Recently, four cases of sepsis attributable to transfusion transmitted Yersinia enterocolitica infection have been reported (23). The cases occurred over a 17 month period. This very rare event seems to be a result of collection of blood from donors who were themselves infected by Y. enterocolitica in association with a very long storage time for the packed red cell units. The latter circumstance apparently permitted significant growth of the bacteria within the blood product. At least two of the donors themselves had histories of gastrointestinal illness during the two weeks prior to transfusion. The major interest in these cases is that this seems to be one of the few bacterial pathogens which have been transmitted as a result of a bacteremic infection in the donor. The frequency of this event is too low to make any recommendations for control, although the Centers for Disease Control are requesting that future cases of this type be reported to them (23).

C. Others

Blood is collected aseptically and when freshly collected, appears to have inherent antibacterial capacity. Consequently, there is little chance of bacterial contamination of donor blood. However, there have been reports of a few cases of bacterial sepsis among recipients of platelet concentrates. In one series, Salmonella sp. infections were thought to be due to collection of blood from transiently bacteremic donors (38). In another series, the infection was due to Staphylococcus epidermidis (16), which was thought to have contaminated the blood unit from the small skin plug which may be produced during venipuncture, even though the site is prepared with an antibacterial scrub. It was considered likely that, in both series, bacterial proliferation occured in the platelet packs during the routine room temperature storage. As a result of these findings, the acceptable storage period for platelets was reduced from 7 to 5 days.

IV. HEPATITIS

Posttransfusion viral hepatitis may be caused by a number of different agents, including cytomegalovirus (73) and, on rare occasions, hepatitis A virus (9,42,53). However, the most frequent causative agent is the NANB virus, although there is evidence that this term may actually apply to a group of two or more viruses with a parenteral transmission mode (15,52,72). Until recently, there has been no specific marker for this virus, so diagnosis and screening have depended upon indirect methods. The reader is referred elsewhere for further details about NANB (27,28). Hepatitis B virus continues to evoke concern and it is clear that a small number of infectious donations continue to be collected, as evidenced by the continued infectivity of pooled plasma products. However, there are continued major improvements in the safety of transfusion, largely as a result of the success of the various measures described below.

A. Donor Selection

The simplest, and most cost effective selection method is to eliminate the collection of blood from unacceptable groups of donors. The two prime examples of this approach have been the elimination of collection activities in prisons and the almost total abandonment of whole blood collections from paid donors. In fact, Allen (4) has shown that recipients of blood collected in penal institutions in the late 1940s and early 1950s had a 1.5% incidence of clinically apparent, acute posttransfusion hepatitis. Current figures, based upon reporting rates for acute posttransfusion hepatitis suggest that the incidence rate among recipients of voluntary blood, is of the order of 0.02%. Even if current reporting rates represent only one third of the true rate of disease (6), the decrease in rates is remarkable. Even prior to the implementation of HBsAg testing, the frequency of reported, acute clinical posttransfusion hepatitis was about 0.2% among recipients of voluntary blood (30). It is also of interest that the frequency of NANB was greater among recipients of non voluntary blood in the Transfusion-Transmitted Viruses study (1) and that, more recently, it has been reported that the prevalence of antibodies to HTLV-1 is about tenfold higher among paid plasma donors than among voluntary donors of whole blood (Lee et al, personal communication).Nevertheless, prospective donors with a history of clinical hepatitis do have significantly increased frequencies of markers of infection, both in early (71) and recent (Tegtmeier, presented at the 1988 meeting of the International Society for Blood Transfusion, London) studies. Indeed, in the latter case, an elevated frequency of the putative marker of NANB hepatitis was found in a population of donors deferred for a history of hepatitis. Since acute infection with HBV and NANB viruses may lead to chronic infection, the measure seems to be of value, although neither its sensitivity nor specificity can be judged.

Other risk factors for HBV infection are routinely sought during the health screening process. In particular, and of considerable current importance, a history, or evidence of, parenteral drug abuse is cause for permanent rejection of a prospective donor. It is increasingly apparent that drug abuse is a major risk factor for infection with HIV and with HTLV I and II (60). Although such infections are found only infrequently in the donor population, former intravenous use of illicit drugs, or a history of sex with such a user, is clearly a risk factor for donor infection, particularly with HTLV 1 (80). Finally, donors are temporarily deferred on the basis of events which are acknowledged to increase risk of infection, at least with respect to infection with HBV.

B. Laboratory Testing

An ideal test would be sensitive, with the capability of detecting all infected or infectious donors and specific, thus avoiding the inappropriate deferral of donors or donations as a result of false positive findings. A distinction must now be drawn between specific and surrogate tests. Specific tests, as their name suggests, define the presence or absence of a definite marker, or footprint of the viral infection in question. Surrogate tests, on the other hand, use some laboratory marker, not necessarily directly related to the virus in question, in order to identify a population of donors considered to be at increased risk of infectivity.

1. Specific testing for HBsAg. Recognition that HBsAg was specifically associated with one form of viral hepatitis led to the relatively rapid implementation of routine HBsAg testing of all blood donations, starting in the early 1970s. Early, gel precipitation methods were rapidly replaced with immunoassays which reached their current levels of sensitivity by the early 1980s. Tests are capable of detecting HBsAg at levels of the order of 1 ng/ml in serum or plasma. Current reporting rates for posttransfusion hepatitis are extremely low and differ little from the incidence of the disease reported in non-transfused populations. It is generally accepted, however, that there may be some residual infectivity, attributable either to collection of blood during the window period between the loss of detectable HBsAg and the development of anti-HBs with resolution of infection, or as a result of very low levels of HBsAg. It has been presumed that the introduction of testing for antibodies to hepatitis B core (anti-HBc) as a surrogate measure for NANB may impact upon this residual infectivity.

Implementation of HBsAg testing was followed by the realization that posttransfusion hepatitis had not been eliminated. This, along with the development of reliable diagnostics for infection with the hepatitis A virus, resulted in the concept of the NANB virus. It was anticipated that a specific test for this virus would become available fairly rapidly. Many candidate markers have been proposed over the years; until recently, all had failed to survive careful testing upon pedigreed panels of samples. Within the last few months, however, there have been strong indications that an antigenic component of a NANB virus has been expressed by genetic engineering and that antibodies to this antigen are strongly associated with NANB hepatitis and may indeed be diagnostic. It is also possible that detection of such antibodies may be used as a screening test for NANB infectivity in donors. At the time of writing, however, there are no primary publications on this promising advance.

2. Use of Surrogate Marker Tests. Extensive and careful prospective studies, performed in the late 1970s established that a high proportion (7 to 12%) of transfused patients developed laboratory signs of liver dysfunction in the absence of evidence of infection with known causative agents of hepatitis (1,5). These events were attributed to infection with NANB virus(es). Although the vast majority of these events were asymptomatic, at least in the acute phase, it became apparent that an alarming proportion of the patients developed prolonged transaminase elevations accompanied by histological evidence of serious liver disease (13). The prospective studies did not succeed in identifying the causative agents, or any specific markers. However, they were able to show an increased prevalence of elevations in alanine aminotransferase among the population of donors whose blood was transfused to the patients who themselves developed transaminase abnormalities. Consequently, it was suggested that populations of donors with ALT elevations were at increased risk of transmitting NANB and that blood should be tested for this marker (1,5). A number of blood centers took an early lead, implementing ALT testing around 1981. Later reanalysis of two of these posttransfusion studies also suggested that the presence of anti-HBc was also a risk factor for transmission of NANB (46,69). Continued concern about the chronic sequelae of NANB infection, plus a strengthened determination to protect blood recipients from infection, led to the widespread adoption of both ALT and anti-HBc testing in the United States.

The adoption of these two tests was expected to result in a reduction in the incidence of posttransfusion NANB of around 60%; such a reduction, at least in reported acute posttransfusion hepatitis, has been reported. However, this has coincided with other changes in the epidemiologic patterns of hepatitis and in the structure of the donor population. Whether or not these changes are a direct result of the implementation of surrogate testing is not known. However, these tests have added a new dimension to laboratory screening, because of the relatively high rates of positive findings. In fact, overall, some 4% to 6% of donations have elevated ALT levels or anti-HBc and must be discarded. This proportion is very much greater than the rejection rate from HBsAg and HIV tests.

C. Results of Measures to Reduce Posttransfusion Hepatitis

It is reasonable to ask how effective these measures have been in reducing the frequency of posttransfusion hepatitis and also to assess the impact upon the donor population and the blood supply. Table 1 shows the rates at which posttransfusion hepatitis was reported to the American Red Cross over the three most recent years (ending in June, 1988) and the rate for a comparable period in 1981-82. These rates represent total collections of 6 to 6.5 million blood units each year. There has clearly been a continuing decline in the frequency of reported disease over the past three years, with the most striking change occurring since the middle of 1987. Interestingly, this decline is also reflected in national data reported to, and investigated by C.D.C. During the period 1983 to 1986, approximately 3500 cases of NANB have been reported annually and in cases where follow up has been completed, 11 to 13% of the cases were epidemiologically associated with transfusion. In contrast, during the calendar year 1987, 3000 cases were reported and the proportion of investigated cases associated with transfusion had dropped to 8%. Similarly, but perhaps more strikingly, over the same period, 22,000 to 26,000 cases of hepatitis B were reported nationally. The proportion of investigated cases associated with transfusion ranged from 3.1 to 4% from 1983 to 1986, but declined to only 1% in 1987.

These declines in reporting rates for posttransfusion hepatitis are consistent and therefore seem unlikely to be a result of changes in reporting efficiency or procedures. It is certainly tempting to attribute the reduced frequencies of posttransfusion disease to widespread implementation of surrogate screening for NANB. It does seem likely that this was a component in the 50% reduction observed over the last year. However, data from C.D.C. Sentinel Counties studies (M. Alter, personal communication), along with the data in Table 1 suggest that this decline was apparent before the impact of testing would be measurable. The fact that homosexual men are at risk for hepatitis B and, perhaps NANB suggests that efforts aimed at reducing the proportion of donors with AIDS risk may well have had a positive impact on posttransfusion hepatitis.

Table 1. Cases of posttransfusion hepatitis reported to the American Red Cross, for one year periods, 1981-1988

Year ending June 30:	Hepatitis B	NANB	Others	All cases
1982	222	434	318	974
1986	166	401	276	843
1987	141	353	190	684
1988	108	194	113	415

V. HIV AND AIDS

A. Donor Selection

Recognition of the potential for transmission of AIDS by transfusion preceded the isolation of the etiologic agent. Consequently, it became critically important to reduce the proportion of potential blood donors with risk factors for AIDS. Starting in 1983, a number of measures were introduced for this purpose and, since that time, there has been a continuing, and apparently successful, series of additions and enhancements to the original programs. Thus, at this time, all donors are extensively educated about risk factors for transmission of HIV and are required to decline to donate if they are at risk. The risk factors include: a history of even one homosexual encounter between males since 1977; intravenous use of drugs of abuse; residence in AIDS endemic areas or travel to such areas if there was a potential for HIV infection; receipt of blood products; a history of prostitution or use of a prostitute, or a histroy of sexual exposure to any person with any of these risk factors. A history of symptoms associated with HIV infection is also sought and donors with these symptoms are not permitted to give blood. The donor must acknowledge, by signature, that he or she is not at risk. In addition, the donor must confidentially specify whether or not the donated unit is safe for transfusion to patients, thus permitting retrieval of potentially unsafe blood which was donated in the face of these restrictions. Another innovative aspect of selection of safe donors was the establishment of alternate test sites, where individuals who wish to know their HIV status may be tested confidentially or anonymously. These test sites were established when the availability of tests was restricted to blood collection establishments and it was feared that people would use blood donation to obtain a test result.

B. Testing for HIV

1. HIV antibody testing. Recognition of the causative agent of AIDS, by Montagnier (12), Gallo (33) and others (49) was accompanied by the development of simple enzyme-linked immunoassay (ELISA) tests for antibodies to HIV (66). Since HIV is a retrovirus, infection is persistent and essentially lifelong. Thus, the presence of antibodies implies the presence of virus and antibody tests may be used to identify potentially infectious blood donations. In May of 1984, virus and cell lines were made available to a group of American companies with the potential to manufacture screening tests and in March 1985, after less than one year, anti-HIV tests were first licensed for use.

ELISA tests for antibodies to HIV almost all have the same basic design, with a solid phase capture reagent bearing HIV antigens and a probe consisting of an enzyme linked antiglobulin reagent. The test sample is serum or plasma and it is generally diluted prior to initiation of the test. Antibodies to HIV, if present, adhere to the solid phase. After suitable washing, the adherent antibodies can be detected with the probe which is itself detected by a suitable chromogenic substrate. An important variation is the use of viral antigens prepared by recombinant DNA technology (17) or by direct synthesis of immunodominant peptides (76). These variant procedures enjoy extensive use in Europe and are apparently at least equivalent to those in use in the United States, where they have not yet been licensed by the Food and Drug Administration.

Screening tests for donated blood are usually conducted according to a simple algorithm. Each sample is tested singly and all those found to be nonreactive are considered to be negative for HIV antibodies. Samples which generate a reactive value are retested in replicate. If either or both of these replicate tests are reactive, the blood unit and all components are recorded as repeatably reactive and must be discarded. The final stage in the algorithm is confirmatory testing in order to determine the extent and nature of donor notification which is required. Careful evaluation clearly suggests that nonrepeatable reactive findings are not associated with increased risk of infectivity (78). Neither does it appear that a repeatably ELISA reactive result with a negative Western blot offers any excess risk of infectivity (36).

While repeatedly reactive ELISA test results are considered to be adequate for a decision to discard a blood unit, they must not be used to notify a donor or other individual that he or she has been infected with HIV. Confirmatory testing must be performed in order to differentiate the true positives from the false positives. Before discussing these supplementary procedures, it is appropriate to comment on the possible reasons for the occurrence of false positives in anti-HIV tests. First, as noted above, the virus which is used to prepare the capture reagent is grown in tissue culture and the mature virus incorporates cell membrane material. This material cannot be fully purified away from the other viral components and, in some cases expresses HLA antigens. Consequently, complementary antibodies, if present in the donor serum, will adhere to the solid phase and generate a repeatable reactive test result. Although this mechanism has been discussed in the literature (45,67), manufacturers of test kits appear to have resolved this problem and there is little evidence that it continues to cause difficulties. However, the final steps of the tests are nonspecific inasmuch as they are designed to detect immunoglobulins. Thus, nonspecifically "sticky" immunoglobulins, or those with an affinity for polystyrene surfaces, will also generate false positive results. Finally, it is clear that there is a more troublesome form of false positive reaction in which there are antibodies which react with polypeptide sequences which are coded by the viral genome, but which do not reflect actual infection with HIV (29); these will be discussed in more detail below.

2. Western blot testing. There are a number of methods available for confirming a reactive ELISA test for anti-HIV, including the use of indirect immunofluorescence (20) and radioimmunoprecipitation (37,64), but the procedure which is most commonly employed in association with blood collection is the Western blot. This procedure permits the identification of antibodies to individual viral polypeptides and careful interpretation of the patterns can result in extremely specific and suitably sensitive identification of truly infected individuals. In addition, the absence of any reactivity in the Western blot is clear evidence that the sample does not come from an individual infected with HIV.

The Western blot procedure (66) involves the lysis of purified HIV and the electrophoresis of the resulting polypeptides on polyacrylamide gel in the presence of SDS. Viral polypeptides migrate at a rate which is inversely proportional to their molecular weight. Once the electrophoresis is complete, the polypeptide pattern is transferred (blotted) to a sheet of nitrocellulose paper. The transfer is accomplished by placing the paper on the face of the acrylamide gel and applying an electric field across the gel. As a result, the band pattern migrates electrophoretically into the nitrocellulose. The nitrocellulose may be dried and cut into narrow strips. In order to perform an assay, the strip is incubated in a dilution of the sample to be tested. Antibodies to viral polypeptides, if present, adhere at positions on the strip which are characteristic. After incubation, the strip is washed and adherent antibodies are detected by means of an appropriately labeled antiglobulin. Usually, the antiglobulin is conjugated to an enzyme and its presence is detected by a suitable color reaction.

The Western blot provides additional information as it permits the separate identification of antibodies to individual viral polypeptides. However, interpretation of a blot is not particularly easy and well-defined criteria must be used in order to read a blot. Currently accepted criteria also require that the blot system must, as a minimum, be able to detect antibodies to the major structural polypeptides of the virus. More specifically, these are the gag products, p17, p24 and the parent p55; the pol products, p31, p51 and p66, and the env products, gp41, gp120 and gp160. The absence of any visible bands in a properly performed blot signifies that the donor or patient is not infected with HIV. Conversely, the presence of certain combinations of bands is regarded as a positive reading and defines infection with HIV for diagnostic purposes and for notification of donors (19).

Unfortunately, patterns other than clear positives or negatives are also seen when samples are subjected to Western blot testing. When a large population of blood donor samples which are repeatedly reactive in anti-HIV ELISA tests are subjected to Western blot, about 10% are positive, about 80% are negative and 10% have indeterminate patterns. Indeterminate patterns cannot be interpreted without further epidemiologic, clinical or virologic information. This is because at least some patterns, notably the presence of isolated reactions to gag peptides, may be due to cross reacting antibodies apparently unassociated with HIV infection (29), or may represent the early stages of seroconversion in an infected individual (64). Among blood donors, it appears that about five of indeterminate patterns may reflect true infection. Indeterminate blot patterns which do not progress towards a full pattern over three to six months are not considered to reflect HIV infection. Thus, the status of the majority of indeterminate patterns can be resolved by testing the donor again, after three to six months.

3. HIV antigen testing. Infection with HIV is not immediately accompanied with the development of detectable levels of antibodies. Nevertheless, there is evidence that HIV infection may be transmitted by blood drawn in the period between infection and antibody detection (79). It has also been shown that sensitive techniques, such as the polymerase chain reaction for HIV genome can, in rare cases, show the presence of HIV some time prior to antibody development or after the apparent disappearance of HIV antibodies (32,82). These findings have generated concern about the sensitivity of ELISA tests for anti-HIV as a donor screening measure. Recently, it has been shown that a soluble antigen with the specificity of the p24 gag polypeptide may be found in the circulation during certain phases of HIV infection. In particular, the antigen is transiently present during the early infection stage, and prior to the development of overt disease (3,35). It has therefore been suggested that blood donors should be tested for the HIV p24 antigen in order to detect early infection. However, limited testing suggests that the antigen tests are positive only for a relatively brief period within a week or two of the appearance of antibodies detectable by conventional tests. It is apparent that relatively few additional samples would be detected: data from Bavaria, for example, showed that there were no samples which were positive only for the HIV antigen when 150,000 successive donations were tested (10). It is also possible that antibody tests of greater sensitivity, particularly for envelope antigens, would be equally effective in detecting this early phase (64).

A number of different HIV antigen tests are available. All are antigen capture procedures, in which the solid phase is antibody to the HIV antigen and the captured analyte is detected by a labeled antibody system. In at least some configurations, the probe is a sequence of antibodies, which presumably increase the sensitivity of the method.

C. Impact of HIV Screening and Testing

In the first year of HIV antibody testing, the American Red Cross identified about 1,600 donations which were confirmed to be positive by Western blot analysis (46,68). A similar number of positives were identified among all other collections. Since it is clear that almost all such donations are infectious for HIV (56,77), screening clearly prevented a significant number of cases of posttransfusion AIDS which would have occurred in the absence of testing. In subsequent years, the total number of detected positives has decreased substantially, largely as a result of the continued elimination of seropositive individuals from the pool of repeat donors. Table 2 outlines these findings. In addition, the data do not suggest that the prevalence of HIV infection among new donors is increasing, neither is there any significant increase in incidence rates among repeat donors. It is also important to note that the frequency of positive test results among donors is much lower than that seen among other tested groups, such as military recruits (18), inner city emergency room patients (11) or randomly selected newborns (40). In addition, the current rate of around 12 per hundred thousand is forty fold lower than would be expected for a random sample of the U.S. population on the assumption that there are around one million infected individuals in the U.S. (21,26). Clearly, donor selection procedures are successful in maintaining a safe and healthy donor population.

Although it is clear that laboratory testing has identified a large number of infectious donations, it is not quite so easy to define the efficacy of the testing. In other words, what proportion of all potentially infectious blood units have been identified? The proportion of AIDS cases attributable to blood transfusion is approximately 3% and this proportion did not change significantly over the three years following the introduction of donor testing. This appears to be a result of the lengthy incubation period for AIDS and the extent of infection which occurred prior to initiation of testing (57). In fact, informal reports from the Centers for Disease Control suggest that, by mid 1988, only about four cases of transfusion-associated AIDS were linked to transfusion which took place after anti-HIV testing started. Allen (personal communication) has compared risk factors in two comparable birth cohorts of infants with AIDS, representing two-year periods before and after testing. In the first group, a total of 78 were shown to have maternal infection as a risk factor and 9 had transfusion as the sole risk factor. In contrast, in an exactly comparable cohort studied after the initiation of donor screening, 258 infants with AIDS had maternal infection as a risk factor, whereas only one (which has not been fully investigated) may have been infected via transfusion (68). Taken together, these reports suggest a meaningful decrease in the frequency of transfusion AIDS as a result of testing.

Table 2. Seroprevalence rates for HIV per 100,000
Red Cross blood donors, by year

Year ending April 30:	First time donors	Repeat donors	All donors
1986	51.7	23.1	27.3
1987	41.8	12.0	16.6
1988	35.6	8.8	12.8

Data selected from 41 Regions.

At the same time, there is certainly evidence that infectious but seronegative blood units are present in the blood supply. Ward and colleagues (79) discussed thirteen blood recipients who had been infected with HIV as a result of receipt of products from 7 donors, most of whom were shown to have engaged in AIDS risk behavior three to four months prior to the donation. This article was accompanied by a worst case estimate suggesting that the frequency of this type of event might be of the order of one infectious product for every 40,000. It is likely that this estimate is unduly high, as it is based upon an overestimate of the incidence of new infections among donors (Dodd, unpublished). Another estimate of the frequency of seronegative but infectious donors has been made by Kleinman (43), who evaluated the frequency of HIV infection among recipients of the last seronegative donation from donors who were subsequently found to be seropositive. Kleinman showed that 50% of recipients of donations made less than six months prior to the positive donation were infected, but none were infected if blood was given more than six moths before a positive finding. The data were interpreted to suggest that, in Los Angeles, about one donation in 67,000 might be infectious but seronegative. Since prevalence rates for Los Angeles are considerably higher than those for the nationwide donor population, this estimate is probably too high for the United States overall.

In summary, it appears reasonable to suggest that the minimum infectivity of the blood supply immediately prior to the start of donor screening for HIV was about 35 per 100,000, which was the seropositivity rate at that time (68). This contrasts with the current rate, which is thought to be of the order of 1 to 2 per 100,000 or better.

VI. HTLV-I AND OTHER RETROVIRUSES

A. Transfusion Associated Retroviral Infection

The human T-lymphotropic retrovirus type I (HTLV-I) was in fact the first human retrovirus to be isolated and was fist described in the early 1980s (58). It is now considered to be the etiologic agent of adult T-cell leukemia/lymphoma, a rare malignancy in the United States, which is, however quite frequent in other parts of the world, notably Japan and parts of the Caribbean. It ia also thought to cause tropical spastic paraperesis (TSP), a neurologic disease (62). There is clear evidence that HTLV-I may be transmitted by transfusion of cellular components of blood, but there is no evidence for cell-free transmission. Unlike HIV, however, infection with HTLV-I only leads to disease in a small number of infected individuals and when such disease occurs, it is after a lengthy incubation period. To date, there have been no cases of HTLV-I disease attributable to transfusion in the United States and there is only presumptive evidence that TSP has resulted from transfusion transmitted infection in Japan.

Studies in the United States suggest that the overall seroprevalence rate for HTLV-I is low and may be confined to rather restricted risk groups (31). There is, however, a remarkably high prevalence of infection among some populations of drug addicts (61). Interpretation of serologic testing for HTLV- I is complicated by the very close structural relationship between this virus and HTLV-II, which has been associated with hairy cell leukemia in a very small number of instances (63). The cross reactivity is such that antibodies to both viruses are detected by tests designed to identify anti-HTLV-I; more complex test such as Western blot and radioimmunoprecipitation also fail to differentiate between infections with these two viruses.

Blood collection agencies in the United States have elected to test all donors of whole blood for antibodies to HTLV-I (and II). The rationale for doing this in the absence of demonstrated disease in blood recipients includes: the commitment to maintain the highest level of safety for the blood supply; the lengthy incubation period for disease; the presence of the virus in the U.S. population; the transmissibility of the virus by transfusion; the desire to avoid transmission of retroviruses, and the wish to interrupt a potential route of transmission into the general population.

B. Testing

ELISA tests for antibodies to HTLV-I were licensed for use in November of 1988 (22). The structure of the tests is almost identical to those designed to detect antibodies to HIV. The capture reagent consists of disrupted purified virus grown in tissue culture and the probe is an enzyme- labeled anti immunoglobulin preparation. Preliminary studies on the donor population have suggested that about 0.025% will be specifically reactive for anti-HTLV-I (31). As with tests for anti-HIV, some proportion of ELISA reactive samples will represent false positives - early data suggest that this figure will be between 0.1 and 1 percent of donor samples tested, depending upon the particular test in use. Thus, confirmatory procedures will be required in order to identify those ELISA positive individuals who are truly infected. Current recommendations require the presence of identifable antibodies to the p24 gag polypeptide and to one or more env glycoproteins in order to specify a true positive. Unfortunately, such findings cannot always be defined solely by use of a Western blot and in cases where the blot shows reactions only to gag proteins, it is necessary to do further testing with radioimmunoprecipitation in order to define the presence or absence of env reactivities (22).

Estimates derived from seroprevalence studies for HTLV-I among donors suggest that, in the absence of screening, about one blood unit in every seven thousand might result in an HTLV-I infection in the recipient (31). Currently, the sensitivities of the test methods have not been defined, but it is reasonable to anticipate that the implementation of testing will reduce this risk to negligible levels.

VII. OTHER VIRUSES

A. Cytomegalovirus

Numerous studies have shown that CMV may be transmitted from seropositive blood donors to seronegative recipients and, in the case of immunocompromised recipients, the outcome may be serious morbidity or even death. Consequently, it is considered appropriate to provide screened, CMV antibody negative blood to selected recipients at risk. However, because the prevalence rates for anti-CMV frequently exceed 50%, such prescreening must necessarily be limited. In general, blood centers distribute about 1% of their products as CMV seronegative units for the support of neonates, but other applications may well increase this proportion significantly.

1. Tests for donor screening. Currently, all laboratory testing of donor blood uses procedures designed to detect all classes of antibody directed against CMV. While there are numerous available methods, including classical reference procedures such as complement fixation and indirect immunofluorescence, the majority of blood center testing is done by particle agglutination or ELISA procedures (70). ELISA procedures are essentially the same as those described for anti-HIV, with a solid phase capture reagent prepared from viral proteins and a labeled antiglobulin probe. Performance and interpretation of these tests is conventional and they are available in bead and microplate formats. Particle agglutination tests for anti-CMV are however, increasingly popular. Passive hemagglutination tests were developed some years ago, but were found to take too long and were somewhat inconsistent. A rapid latex agglutination procedure is now available; it may be completed in less than 10 minutes. The procedure is simple, rapid and has comparable sensitivity and specificity to other methods (70). It can be used for rapid screening both of collected products and, in some situations, prior to collection.

Although the prevalence rate for CMV antibodies in donor populations is of the order of 50%, review of the outcome of transfusion of seropositive units suggests that only a small proportion of them are actually infectious. It would be desirable to use a test which had the capability of detecting only those units which are infectious. Two candidate systems have been described: tests for IgM antibodies to CMV (47) and tests for antibodies to the so called early antigens of CMV (48). ELISA procedures to detect IgM anti-CMV do not appear to be readily available in the United States, although some procedures have been developed in Europe. Lamberson and colleagues used an indirect immunofluorescence assay to show that, in a population of 1535 blood donors, 38% were reactive for CMV total antibodies in ELISA and 6% were reactive for IgM antibodies. In studies to determine the potential value of donor screening for IgM anti-CMV, the incidence of CMV among neonatal recipients of unscreened blood was 3.2% whereas the incidence was 0.7% among infants receiving only IgM anti-CMV nonreactive blood, representing a single case in which the implicated unit did have a positive ELISA test and detectable levels of IgM anti-CMV by another test. These studies have not been confirmed at this time and it may be too early to make a specific recommendation about this approach.

Another study by Lamberson and his associates (48) compared the results of various serologic tests with viral isolation from urine among 500 donors. Three of the donors (0.6%) were viruric; all three were positive in an IFA test for antibodies to CMV early antigen and in conventional ELISA and latex agglutination tests. One of the viruric donors was identified by an IFA, and two by an ELISA, for IgM anti-CMV. Among all donors, 35% were positive by conventional ELISA, 16% were positive in the test for antibodies to early antigens and 3% were positive by the ELISA for IgM anti-CMV. It was suggested that tests for antibodies to the early antigens might have some value in identifying infectious donors. Again, it is too early to make any specific recommendations.

2. Current situation. The major blood banking organizations in the United States have recognized the risk of morbidity or mortality among CMV seronegative, immunocompromised recipients of CMV seropositive blood. The benefit of using screened, CMV seronegative blood for transfusion of seronegative, low birthweight infants (i.e. less than 1250 grams) has been clearly established and this procedure is recommended. Clearly, it would also be desirable to transfuse seronegative blood to a seronegative pregnant woman. There has been some concern about the indiscriminate use of seronegative products in seropositive infants however, since dilutional effects may result in increased morbidity (1073). The benefits of using seronegative blood products for other patient groups are not so clear and must be weighed against the availability of an adequate supply. Organ transplant recipients are at risk of reactivation of intrinsic infection and may also be infected via the transplanted organ. It is possible that CMV infection among seronegative recipients of seronegative organs may be controlled by the use of CMV immune globulin, provided granulocytes are not transfused (81). However, recipients of liver transplants may warrant special attention. In addition, bone marrow recipients are at extraordinarily high risk of death resulting from CMV infection and it is considered appropriate to use granulocytes from CMV seronegative donors for these patients (70). It has also been suggested that AIDS patients should be considered at risk of transfusion associated CMV disease (74). It is important to note that, on the other side of the coin, there is no evidence that immunocompetent recipients are at any significant risk of disease from CMV positive transfusions and that the process of testing blood for anti-CMV should be considered more in the light of a compatibility assessment to match an appropriate product to the special needs of a selected patient.

B. Creutzfeld-Jacob Disease

Concern about the potential for transmission of other viruses has been discussed from time to time, but there seem to be few indications for specific precautions. Of interest, however, was the recent finding of seven cases of Creutzfeldt-Jacob disease among the approximately 12,000 recipients of human- derived pituitary growth hormone. The agent of this disease appears to have been transmitted by transplantation of human tissues and has apparently been experimentaly transmitted to rodents from human blood. Consequently, the FDA has required that individuals who have been treated with human pituitary growth hormone be excluded from donating blood (41).

C. The B19 parvovirus

One other virus which is documented as having been transmitted by transfusion is the B19 parvovirus, the causative agent of "fifth disease". However, infection with this virus does not generally result in serious symptoms, other than in individuals with certain forms of anemia and, perhaps in pregnant women (7,25). There seems to be little need for any specific measures to protect blood recipients against infection with this agent.

VIII. SUMMARY AND CONCLUSIONS

Blood, like any other medication, carries a risk of adverse effects. The objective of reducing this risk to zero may not be attainable. However, it is important to recognize that considerable and increasing effort is being expended to make this risk as low as can be attained.

Table 3. Estimates of the frequency of infection or disease from blood transfusion: number of donations per event

Agent	Reported Disease	Estimated Infection
HBV	40,000	300
NANB	10,000	15 - 125
HIV	NA	40,000 - 200,000
HTLV I/II	NA	7,000

Current estimates of the chance of being infected with blood transmissible pathogens, along with estimates of the frequency of reported clinical disease, are presented in Table 3. Most concern has been directed to the risk of HIV infection and it should be noted that this risk is approximately equivalent to the risk of suffering a fatal hemolytic transfusion reaction - one in 100,000. It is also instructive to relate these risk levels to those of other common medical or everyday activities.   There is a 1 in 4350 chance of dying each year as a result of pregnancy (U.K. data), a 1 in 12500 chance of dying of leukemia, a 1 in 5900 chance of dying as a result of driving an automobile (U.K.).   It is also of interest to note that the voluntarily accepted lifetime risk of dying as a result of smoking 20 cigarettes a day is 1 in 200 (84).

Methods used to select donor populations and to screen individual donors sharply reduce the frequency of infectious disease markers to levels well below those found in random population samples. In addition, the implementation of specific and surrogate laboratory tests identifies almost all of the residual infectious units. Improvements are anticipated, particularly with the advent of a specific test for antibodies associated with NANB hepatitis. Active research on depletion of viruses in red cell products, or the application of photoinactivation procedures offers hope for the final elimination of the very few remaining infectious blood donations.

IX. REFERENCES

1. Aach, R.D., W. Szmuness, J.W. Mosley, F.B. Hollinger, R.A. Kahn, C.E. Stevens, V.M. Edwards, and J. Werch. 1981. Serum alanine aminotransferase of donors in relation to the risk of non-A, non-B hepatitis in recipients. The Transfusion-Transmitted Viruses Study. N Eng J Med 304:989-994.

2. Ablashi, D.V., S.Z. Salahuddin, S.F. Josephs, F. Imam, P. Lusso, R.C. Gallo, C. Hung, J. Lemp, and P.D. Markham. 1987. HBLV (or HHV-6) in human cell lines. Nature 329:207-207.

3. Allain, J.-P., Y. Laurian, D.A. Paul, D. Senn, and Members of the AIDS-Haemophilia French Study Group. 1986. Serological markers in early stages of human immunodeficiency virus infection in haemophiliacs. Lancet 2:1233-1236.

4. Allen, J.G. 1972. The epidemiology of posttransfusion hepatitis. Basic blood and plasma tabulations. p.1-335. Commonwealth Fund, Stanford.

5. Alter, H.J., R.H. Purcell, P.V. Holland, D.W. Alling, and D.E. Koziol. 1981. Donor transaminase and recipient hepatitis. Impact on blood transfusion services. JAMA 246:630-634.

6. Alter, M.J., A. Mares, S.C. Hadler, and J.E. Maynard. 1987. The effect of underreporting on the apparent incidence and epidemiology of acute viral hepatitis. Amer J of Epid 125:133-139.

7. Anderson, M.J. 1987. Human parvovirus infections. J of Vir Meth 17:175-181.

8. AuBuchon, J.A., and R.Y. Dodd. 1988. Analysis of the relative safety of autologous blood units available for transfusion to homologous recipients. Transfusion 28:403-405.

9. Azimi, P.H., R.R. Roberto, J. Guralnick, T. Livermore, S. Hoag, S. Hagens, and N. Lugo. 1986. Transfusion-acquired hepatitis A in a premature infant with secondary nosocomial spread in an intensive care nursery. Amer J Dis Children 140:23-27.

10. Backer, U., F. Weinauer, G. Gathof, and J. Eberle. 1987. HIV antigen screening in blood donors [letter]. Lancet 2:1213-1214.

11. Baker, J.L., G.D. Kelen, K.T. Sivertson, and T.C. Quinn. 1987. Unsuspected human immunodeficiency virus in critically ill emergency patients. JAMA 257:2609-2611.

12. Barre-Sinoussi, F., J.-C. Chermann, F. Rey, M.T. Nugeyre, S. Chamaret, J. Gruest, C.

Dauguet, C. Axler-Blin, F. Brun-Vezinet, C. Rouzioux, W. Rozenbaum, and L. Montagnier. 1983. Isolation of a T-lymphotropic retrovirus from a patient at risk for acquired immune deficiency syndrome (AIDS). Science 220:868-871.

13. Berman, M., A.J. Alter, K.G. Ishak, R.H. Purcell, and A.E. Jones. 1979. The chronic sequelae of non-A, non-B hepatitis. Ann of Intern Med 91:1-6.

14. Blattner, W.A., D.W. Blayney, M.R. Guroff, M.G. Sarngadharan, V.S. Kalyanaraman, P.S. Sarin, E.S. Jaffe, and R.C. Gallo. 1983. Epidemiology of Human T-Cell Leukemia/Lymphoma Virus. J of Inf Dis 147:406-416.

15. Bradley, D.W., J.E. Maynard, H. Popper, E.H. Cook, J.W. Ebert, K.A. McCaustland, C.A. Schable, and H.A. Fields. 1983. Posttransfusion non-A, non-B hepatitis: Physicochemical properties of two distinct agents. J of Inf Dis 148:254-265.

16. Braine, H.G., T.S. Kickler, P. Charache, P.M. Ness, J. Davis, C. Reichart, and A.K. Fuller. 1986. Bacterial sepsis secondary to platelet transfusion: an adverse effect of extended storage at room temperature. Transfusion 26:391-393.

17. Burke, D.S., B.L. Brandt, R.R. Redfield, T.H. Lee, R.M. Thorn, G.A. Beltz, and C.H. Hung. 1987. Diagnosis of human immunodeficiency virus infection by immunoassay using a molecularly cloned and expressed virus envelope polypeptide. Comparison to Western blot on 2707 consecutive serum samples. Ann Intern. Med 106:671-676.

18. Burke, D.S., J.F. Brundage, J.R. Herbold, W. Berner, L.I. Gardner, J.D. Gunzenhauser, J. Voskovitch, and R.R. Redfield. 1987. Human immunodeficiency virus infections among civilian applicants for United States military service, October 1985 to March 1986. Demographic factors associated with seropositivity. N Eng J Med 317:131-136.

19. Carlson, J.R. 1988. Serological diagnosis of human immunodeficiency virus infection by Western blot testing. JAMA 260:674-679.

20. Carlson, J.R., J. Yee, S.H. Hinrichs, M.L. Bryant, M.B. Gardner, and N.C. Pedersen. 1987. Comparison of indirect immunofluorescence and Western blot for detection of anti-human immunodeficiency virus antibodies. J Clin. Microbiol 25:494-497.

21. CDC 1987. Human immunodeficiency virus infection in the United States: A review of current knowledge. MMWR 26 (Supplement S6):1-48.

22. CDC 1988. Licensure of screening tests for antibody to human T-lymphotropic virus type I. MMWR 37:736-745.

23. CDC 1988. Yersinia enterocolitica bacteremia and endotoxin shock associated with red blood cell transfusion-United States, 1987-1988. MMWR 37:577-578.

24. CDC 1988. Safety of therapeutic products used for hemophilia patients. MMWR 37:441-450.

25. CDC 1989. Risks associated with human parvovirus B19 infection. MMWR 38:81-97.

26. Curran, J.W., H.W. Jaffe, A.M. Hardy, W.M. Morgan, R.M. Selik, and T.J. Dondero. 1988. Epidemiology of HIV infection and AIDS in the United States. Science 239:610-616.

27. Dienstag, J.L. 1983. Non-A, non-B hepatitis. II. Experimental transmission, putatitve virus agents and markers, and prevention. Gastroenterol 85:743-768.

28. Dienstag, J.L. 1983. Non-A, non-B hepatitis. I. Recognition, epidemiology, and clinical features. Gastroenterol 85:439-462.

29. Dock, N.L., H.V. Lamberson,Jr., T.A. O'Brien, D.E. Tribe, S.S. Alexander, and B.J. Poiesz. 1988. Evaluation of atypical human immunodeficiency virus immunoblot reactivity in blood donors. Transfusion 28:412-418.

30. Dodd, R.Y. 1985. Donor screening and epidemiology, p.389-405. In R.Y. Dodd, and L.F. Barker (eds.), Infection, Immunity, and Blood Transfusion. Alan R. Liss, New York.

31. Dodd, R.Y., and N. Nath. 1987. Increased risk for lethal forms of liver disease among HBsAg-positive blood donors in the U.S. J of Vir Meth 17:81-94.

32. Farzadegan, H., M.A. Polis, S.M. Wolinsky, C.R. Rinaldo, J.J. Sninsky, S. Kwok, R.L. Griffith, R.A. Kaslow, J.P. Phair, F. Polk, and A.J. Saah. 1988. Loss of human immunodeficiency virus type 1 (HIV-1) antibodies with evidence of viral infection in asymptomatic homosexual men. Ann of Intern Med 108:785-790.

33. Gallo, R.C., S.Z. Salahuddin, M. Popovic, G.M. Shearer, M. Kaplan, B.F. Haynes, T.J. Palker, R. Redfield, J. Oleske, B. Safai, G. White, P. Foster, and P.D. Markham. 1984. Frequent detection and isolation of cytopathic retroviruses (HTLV-III) from patients with AIDS and at risk for AIDS. Science 224:500-503.

34. Gomperts, E.D. 1986. Procedures for the inactivation of viruses in clotting factor concentrates. Am J Hematol. 23:295-305.

35. Goudsmit, J., F. DeWolf, D.A. Paul, L.G. Epstein, J.M.A. Lange, W.J.A. Krone, H. Speelman, E.C. Wolters, J. van der Noorda, J.M. Oleske, H.J. van der Helm, and R.A. Coutinho. 1986. Expression of human immunodeficiency virus antigen (HIV-Ag) in serum and cerebrospinal fluid during acute and chronic infection. Lancet 2:177-180.

36. Grindon, A.J., S.E. Critchley, and J.W. Ward. 1988. Risk of HIV infection in recipients of untested blood from donors now anti-HIV-positive. Transfusion 28:419-421.

37. Handsfield, H.H., M. Wandell, L. Goldstein, and K. Shriver. 1987. Screening and diagnostic performance of enzyme immunoassay for antibody to lymphadenopathy-associated virus. J. Clin. Microbiol. 25:879-884.

38. Heal, J.M., M.E. Jones, J. Forey, A. Chaudry, and R.L. Stricoff. 1987. Fatal Salmonella septicemia after platelet transfusion. Transfusion 27:2-5.

39. Henle, W., and G. Henle. 1985. Epstein-Barr virus and blood transfusions, p.201-209. In R.Y. Dodd, and L.F. Barker (eds.), Infection, Immunity, and Blood Transfusion. Alan R. Liss, New York.

40. Hoff, R., V.P. Berardi, B.J. Weiblen, L. Mahoney-Trout, M.L. Mitchell, and G.F. Grady. 1988. Seroprevalence of human immunodeficiency virus among childbearing women. Estimation by testing samples of blood from newborns. N. Engl. J Med. 318:525-530.

41. Holland, P.V. 1988. Why a new standard to prevent Creutzfeldt-Jakob disease. Transfusion 28:293-294.

42. Hollinger, F.B., N.C. Khan, P.E. Oefinger, D.H. Yawn, A.C. Schmulen, G.R. Dreesman, and J.L. Melnick. 1983. Posttransfusion hepatitis type A. JAMA 250:2313-2317.

43. Kleinman, S., and K. Secord. 1988. Risk of human immonodeficiency virus (HIV) transmission by anti-HIV negative blood. Estimates using the lookback methodology. Transfusion 28:499-501.

44. Koziol, D.E., P.V. Holland, D.W. Alling, J.C. Melpolder, R.E. Solomon, R.H. Purcell, L.M. Hudson, F.J. Shoup, H. Krakauer, and H.J. Alter. 1986. Antibody to hepatitis B core antigen as a paradoxical marker for non-A, non-B hepatitis agents in donated blood. Ann of Intern Med 104:488-495.

45. Kuhnl, P., S. Seidl, and G. Holzberger. 1985. HLA Dr4 antibodies cause positive HTLV III antibody ELISA results. Lancet 1:1222-1223.

46. Kuritsky, J.N., S.C. Rastogi, G.A. Faich, J.B. Schorr, J.E. Menitove, R.W. Reilly, and

J.R. Bove. 1986. Results of nationwide screening of blood and plasma for antibodies to human T-cell lymphotropic III virus, type III. Transfusion 26:205-207.

47. Lamberson, H.V., J.A. McMillan, L.B. Weiner, M.L. Williams, D.A. Clark, C.A. McMahon, E.B. Lentz, A.P. Higgins, and N.L. Dock. 1988. Prevention of transfusion-associated cytomegalovirus (CMV) infection in neonates by screening blood donors for IgM to CMV. J of Inf Dis 157:820-823.

48. Lentz, E.B., N.L. Dock, C.A. McMahon, S.R. Fiesthumel, C.B. Arnold, and H.V. Lamberson. 1988. Detection of antibody to CMV-induced early antigens and comparison with four serologic assays and presence of viruria in blood donors. J of Clin Microbiol 26:133-135.
49. Levy, J.A., A.D. Hoffman, S.M. Kramer, J.A. Landis, J.M. Shimabukuro, and L.S. Oshiro. 1984. Isolation of lymphocytoathic retroviruses from San Francisco patients with AIDS. Science 225:840-842.

50. Mannucci, P.M., and M. Colombo. 1988. Virucidal treatment of clotting factor concentrates. Lancet 2:782-784.

51. Minamoto, G.Y., J.W.M. Gold, D.A. Scheinberg, W.D. Hardy, N. Chein, E. Zuckerman, L. Reich, K. Dietz, T. Gee, J. Hoffer, K. Mayer, J. Gabrilove, B. Clarkson, and D. Armstrong. 1988. Infection with human T-cell leukemia virus type I in patients with leukemia. N Eng J Med 318:

52. Mosley, J.W., A.G. Redeker, S.M. Feinstone, and R.H. Purcell. 1977. Multiple hepatitis viruses in multiple attacks of acute viral hepatitis. N Eng J Med 296:75-78.

53. Noble, R.C., M.A. Kane, S.A. Reeves, and I. Roeckel. 1984. Posttransfusion hepatitis A in a neonatal intensive care unit. JAMA 252:2711-2715.

54. Okochi, K., and H. Sato. 1986. Transmission of adult T-cell leukemia virus (HTLV-1) through blood transfusion and its prevention. AIDS Res 2:S157-S161.

55. Okochi, K., H. Sato, and Y., Hinuma. 1984. A retrospective study on transmission of adult T-cell leukemia virus by blood transfusion: Seroconversion in recipients. Vox Sang 46:245-253.

56. Peterman, T.A., K.-J. Lui, D.N. Lawrence, and J.R. Allen. 1987. Estimating the risks of transfusion-associated acquired immune deficiency syndrome and human immunodeficiency virus infection. Transfusion 27:371-374.

57. Peterman, T.A., R.L. Stoneburner, J.R. Allen, H.W. Jaffe, and J.W. Curran. 1988. Risk of human immunodeficiency virus transmission from heterosexual adults with transfusion-associated infections. JAMA 259:55-58.

58. Poiesz 1980. Detection and isolation of type C retroviruses particles from fresh and cultured lymphocytes of a patient with cutaneous T-cell lymphoma. Proc Natl Acad Sci Vol 77, No 12:7415-7419.

59. Popovsky, M.A., L.E. Lindberg, A.L. Syrek, and P.L. Page. 1988. Prevalence of Babesia antibody in a selected blood donor population. Transfusion 28:59-61.

60. Robert-Guroff, M., S. Weiss, J. Giron, A.M. Jennings, H.M. Ginzburg, I.B. Margolis, W.A. Blattner, and R.C. Gallo. 1986. Prevalence of antibodies to HTLV-1, -II, and -III in intravenous drug abusers from an AIDS endemic region. JAMA 255:3133-3137.

61. Robert-Guroff, M., S.H. Weiss, J.A. Giron, A.M. Jennings, H.M. Ginzburg, I.B. Margolis, W.A. Blattner, and R.C. Gallo. 1986. Prevalence of antibodies to HTLV-I, -II, and -III in intravenous drug abusers from an AIDS endemic region. JAMA 255:3133-3137.

62. Roman, G.C. 1987. Retrovirus-associated myelopathies. Arch Neurol 44:659-663.

63. Rosenblatt, J.D., I.S.Y. Chen, and W. Wachsman. 1988. Infection with HTLV-I and HTLV-II: Evolving concepts. Semin. Hematol. 25:230-246.

64. Saah, A.J., H. Farzadegan, R. Fox, P. Nishanian, C.R. Rinaldo, J.P. Phair, J.L. Fahey, T.H. Lee, B.F. Polk, and The Multicenter AIDS Cohort Study. 1987. Detection of early antibodies in HIV infection by enzyme-linked immunosorbent assay, western blot, and radioimmunoprecipitation. J of Clin Microbiol 25:1605-1610.

65. Salahuddin, S.Z., D.V. Ablashi, P.D. Markham, S.F. Josephs, S. Sturzenegger, M. Kaplan, G. Halligan, P. Biberfeld, F.W. Staal, B. Kramarsky, and R.C> Gallo. 1986. Isolation of a new virus, HBLV, in patients with lymphoproliferative disorders. Science 234:596-600.

66. Sarngadharan, M.G., M. Popovic, L. Bruch, J. Schupbach, and R.C. Gallo. 1984. Antibodies reactive with human T-lymphotropic retroviruses (HTLV-III) in the serum of patients with AIDS. Science 224:506-508.

67. Sayers, M.H., P.G. Beatty, and J.H. Hansen. 1986. HLA antibodies as a cause of false-positive reactions in screening enzyme immunoassays for antibodies to human T-lymphotropic virus type III. Transfusion 26:113-115.

68. Schorr, J.B., A. Berkowitz, P.D. Cumming, A.J. Katz, and S.G. Sandler. 1985. Prevalence of HTLV-III antibody in American blood donors. N Eng J Med 313:384-385.

69. Stevens, C.E., R.D. Aach, F.B. Hollinger, J.W. Mosley, W. Szmuness, R. Kahn, J. Werch, and V. Edwards. 1984. Hepatitis B virus antibody in blood donors and the occurrence of Non-A, non-B hepatitis in transfusion recipients. An analysis of the Transmission-Transmitted Viruses Study. Ann of Intern Med 101:733-738.

70. Swenson, S. 1987. Syphilis serology, cytomegalovirus testing and alanine aminotransferase testing, p.51-90. In M.R. Dixon, and S.S. Ellisor (eds.), Selection of Methods and Instruments for Blood Banks. American Association of Blood Banks, Arlington, VA.

71. Tabor, E., J.H. Hoofnagle, L.F. Barker, G. Tamondong-Pineda, N. Nath, L.A. Smallwood, and R.J. Gerety. 1981. Antibody to hepatitis B core antigen in blood donors with a history of hepatitis. Transfusion 21:366-371.

72. Tabor, E., D.R. Jackson, Z. Schaff, P.M. Blatt, and R.J. Gerety. 1984. Additional evidence for more than one agent of human non-A, non-B hepatitis. Transmission and passage studies in chimpanzees. Transfusion 24:224-230.

73. Tegtmeier, G.E. 1985. Cytomegalovirus and blood transfusion, p.175-199. In R.Y. Dodd, and L.F. Barker (eds.), Infection, Immunity, and blood transfusion. Alan R. Liss, New York.

74. Tegtmeier, G.E. 1987. Blood transfusion and the transmission of cytomegalovirus, p.87-118. In S.B. Moore (ed.), Transfusion-Transmitted Viral Diseases. American Association of Blood Banks, Arlington, VA.

75. Walker, R.H. 1965. The disposition of STS reactive blood in a transfusion service. Transfusion 5:452-456.

76. Wang, J.J.G., S. Steel, R. Wisniewolski, and C.Y. Wang. 1986. Detection of antibodies to human T-lymphotropic virus type III by using a synthetic peptide of 21 amino acid residues corresponding to a highly antigenic segment of gp41 envelope protein. Proc Natl Acad Sci USA 83:6159-6163.

77. Ward, J.W., D.A. Deppe, S. Samson, H. Perkins, P. Holland, L. Fernando, P.M. Feorino, P. Thompson, S. Kleinman, and J.R. Allen. 1987. Risk of human immunodeficiency virus infection from blood donors who later developed the acquired immunodeficiency syndrome. Ann of Intern Med 106:61-62.

78. Ward, J.W., A.J. Grindon, P.M. Feorino, C. Schable, M. Parvin, and J.R. Allen. 1986. Laboratory and epidemiologic evaluation of an enzyme immunoassay for antibodies to HTLV-III. JAMA 256:357-361.

79. Ward, J.W., S.D. Holmberg, J.R. Allen, D.L. Cohn, S.E. Critchley, S.H. Kleinman, B.A. Lenes, O. Ravenholt, J.R. Davis, M.G. Quinn, and a.l. et. 1988. Transmission of human immunodeficiency virus (HIV) by blood transfusions screened as negative for HIV antibody. N. Engl. J. Med. 318:473-478.

80. Williams, A.E., C.T. Fang, D.J. Slamon, B.J. Poiesz, S.G. Sandler, W.F. Darr, G. Shulman, E.I. McGowan, D.K. Douglas, R.J. Bowman, F. Peetoom, S.H. Kleinman, B. Lenes, and R.Y. Dodd. 1988. Seroprevalence and epidemiological correlates of HTLV-1 infection in U.S. blood donors. Science 240:643-646.

81. Winston, D.J., W.G. Ho, L. Cheng-Hsien, K. Bartoni, M.D. Budinger, R.P. Gale, and R.E. Champlin. 1987. Intravenous immune globulin for prevention of cytomegalovirus infection and interstitial pneumonia after bone marrow transplantation. Ann of Intern Med 106:12-18.

82. Wolinsky, S., C. Rinaldo, H. Farzadegan, P. Gupta, R. Kaslow, D. Immagawa, J. Chmiel, J., Phair, J. Kwok, and J. Sninsky. 1988. Polymerase chain reaction (PCR) detection of HIV provirus before HIV seroconversion. Abstracts: IV International Conference on AIDS. 1:137.(Abstract)

83. Yeager 1983. Sequelae of maternally derived cytomegalovirus infections in premature infants. J of Pediatrics. 102:918-922.

84. Zuck, T.F. 1987. Greetings - a final look back with comments about a policy of a zero-risk blood supply. Transfusion 27:447-448.

CDC, NCCLS, AND OSHA GUIDELINES FOR UNIVERSAL PRECAUTIONS:

WHO IS RIGHT AND ARE THE GUIDELINES PRACTICAL?

Stanley Bauer

Department of Pathology
The Bronx-Lebanon Hospital Center
Bronx, New York, and
The National Committee for Clinical Laboratory
Standards
Villanova, Pennsylvania

INTRODUCTION

Universal Precautions as recommended by the Centers for Disease Control (CDC) in 1987 (1) and updated in 1988 (2) have been widely adopted and form an integral part of the regulatory activity of the Occupational Safety and Health Administration (OSHA) and state regulatory agencies. Universal precautions as promulgated by the CDC imply the use of Blood and Body Fluid Isolation for all patients regardless of any lack of evidence of an infection with hepatitis B virus (HBV), human immunodeficiency virus (HIV), or other blood-borne infection. The implementation of universal precautions has raised many issues which have not been officially addressed by the CDC or OSHA.

Among these unanswered issues are: 1) No variation from recommendations is explicitly granted for local differences in prevalence, although a suggestion of an acceptable modification of universal precautions for voluntary blood-donation sites and routine phlebotomy has been put forth, 2) Practices such as Body Substance Isolation are not commented upon, but may offer valuable modifications, 3) Whereas workers have a legal "Right-To-Know" of hazards in the work-place, the infectious status of a patient or specimen may not be made known to them in an effort to protect the confidentiality of the patient, 4) Under universal precautions, if a health-care worker is accidentally exposed to HIV infectious blood or specified body fluids in a clinical setting, the informed consent of the patient is to be obtained prior to testing the patient's blood for HIV antibodies, whereas, under the revised Agent Summary Statement for HIV, if a laboratory worker is exposed, the source laboratory specimen should be tested for HIV antibodies, antigen, or virus without patient consent, 5) The magnitude of the risk to an immuno-compromised worker is not defined and practices to protect such workers are not delineated, and 6) How universal should universal precautions be?

BACKGROUND

"Universal Blood and Body Fluid Precautions" or "Universal Precautions" were recommended by the CDC in 1987 (1) and updated in 1988 (2). Prior to 1987, the CDC had recommended category-specific and/or disease-specific isolation systems (3). Blood and body fluid isolation was to be used when a patient was known or suspected to be infected with bloodborne pathogens. Included in the category-specific diseases requiring blood and body fluid isolation are:

1. Acquired immunodeficiency syndrome (AIDS)

2. Hepatitis B (including HBsAg carrier)

3. Hepatitis, non-A, non-B

4. Syphilis (with skin and mucus membrane lesions)

5. Creutzfeldt-Jacob disease

6. Several protozoal and viral diseases

The specific practices recommended under blood and body fluid isolation are intended to prevent the infection of health-care workers and the transmission of infection to other patients.

Subsequent to the first description by the CDC (4) of the syndrome which was later to be known as AIDS, the CDC published a number of recommendations for the protection of health-care workers from occupationally acquired HIV (5). Separate documents addressed clinical and laboratory staffs, health-care workers and allied professionals, risks in the workplace, and risks associated with invasive procedures.

The CDC updated all of their previous recommendations with the publication of expanded recommendations for the prevention of HIV transmission in health-care settings in August 1987 (1). The CDC stressed the need to treat all patients as potentially infected with HIV and/or other bloodborne pathogens and to consider blood and body fluids of all patients as potentially infective. This required the utilization of blood and body fluid precautions for all patients. Thus, "Universal Precautions" was added to our lexicon. The elements of universal precautions are formulated to address the needs of the clinical health-care worker. The application of universal precautions in the hospital or research laboratory was not addressed in depth. The recommendation to utilize universal precautions was widely publicized and adopted.

In June, 1987 the National Committee for Clinical laboratory Standards (NCCLS), a voluntary consensus standards organization, anticipated the need for an authoritative, current, bench-level procedure manual which could provide guidance in implementing practices which would protect the hospital clinical laboratory worker from bloodborne infections and would serve as a source of information which was not readily available to the clinical laboratory community. In October, 1987 the NCCLS published a Proposed Guideline entitled "Protection of Laboratory Workers from Infectious Disease Transmitted by Blood and Tissue" (6). The Guideline is the product of the NCCLS Task Force on Protection of Laboratory Workers which met in August, 1987. The task force was composed of representatives of many professional organizations, manufacturers of in-vitro diagnostic products, and governmental agencies. Although many diseases are transmissible by blood, body fluids, and tissue, the guideline focuses on HIV and HBV because these pose a risk of infection which is both common and grave. The Guideline addresses many issues concerning the implementation of a rational system of precautions for use in the clinical laboratory and in anatomic pathology. It is consistent with and supplemental to the general recommendations of the CDC.

In October, 1987 the Department of Labor and the Department of Health and Human Services issued a "Joint Advisory Notice" giving recommendations for protection against occupational exposure to HBV and HIV (7). In January, 1988 the Occupational Safety and Health Administration (OSHA) Office of Health Compliance Assistance issued an instruction for inspection procedures to ensure that health-care facilities were complying with all pertinent OSHA and CDC recommendations including those in the Joint Advisory Notice (8).

Many questions were raised by the introduction of universal precautions to widely varying health-care institutions. As enunciated by the CDC no variations in universal precautions are permitted to institutions serving populations which vary from the extremes of the high prevalence inner-city risk populations of New York and San Francisco to low prevalence affluent suburban and rural communities. The experience of the first year of universal precautions led to an updated set of recommendations by the CDC (2) which addressed some of the earlier unanswered questions. However, CDC still recommends that universal precautions as defined by CDC be utilized for all patients in all health-care settings. In a similar vein, OSHA issued a revised set of instructions for inspection procedures in August, 1988 (9) stating that it will enforce the CDC recommendations in all heath-care delivery settings. No provision is made for variations among institutions in high prevalence inner city locations versus those in low prevalence areas.

In accordance with NCCLS policy of producing consensus documents, the NCCLS Proposed Guideline was widely distributed and a large number of comments were received from a diverse mixture of users. The NCCLS guideline was revised by a subcommittee of the Area Committee on Microbiology and published as a Tentative Guideline in January 1989 (10). In the revised guideline the NCCLS recognizes that the risk of HIV infection varies widely in different population centers in the U.S. Accordingly, it recommends that local circumstances may dictate a modification of the general recommendations either to be more or less rigorous as circumstances dictate. It is recognized that the prevalence of HBV is more uniform in the population, and HBV vaccination is recommended for all high-risk groups of health-care workers.

DEFINING THE RISK

The Nature of the Risk

The source of the risk of nosocomial HIV or HBV infection to the health-care worker is the routine exposure to blood, body fluids, and tissues of patients. The risk of nosocomial infection cannot be avoided since contact with patients, blood, body fluids, or laboratory specimens is integral to the delivery of health-care. Exposure may take place at the bedside, in any area where invasive procedures are carried out, in outpatient facilities, in the laboratory, in the morgue, in research laboratories, and in the home.

If currently recommended safety precautions are followed, there should be no occupational hazard to the health-care worker baring any accidental exposure. It is the potential for accidents, and the certainty that they will occur, which must be considered in implementing a protection program.

It is necessary to examine each individual procedure practiced in the health-care setting to detect potential areas for accidents. Once identified, standard operating procedures (SOPs) should be developed, and are required by OSHA, to minimize the probability and magnitude of accidents. The SOPs will require equipment, supplies, and increased time which may add substantially to the cost of health-care delivery. There is a need to develop an in-depth educational program to train workers in safe practices, to consistently reinforce this education, to monitor the implementation of the safety practices, and to counsel and retrain workers if infractions are detected.

The Magnitude of the Risk

HIV has been found to be transmitted by routes similar to HBV (1), and thus, the precautions used to prevent nosocomial infection with HBV serve well as a model for HIV. HIV has been isolated from blood, semen, vaginal secretions, saliva, tears, breast milk, cerebrospinal fluid, amniotic fluid, alveolar fluid, and urine (1,11). However, only blood, semen, vaginal secretions, and breast milk have been implicated in transmission (1). HIV appears to be less easily transmitted through accidents, such as accidental needlesticks, than does HBV (1). The risk of HBV infection after accidental needlestick has been reported to be between 6% and 30% (12,13) while the seroconversion rate after needlestick exposure to HIV infective blood is estimated to be less than 1% (1,14,15). Long-term follow-up of large numbers of HIV-exposed workers is needed to refine this estimate.

The magnitude of the risk is determined by the nature of the procedure being carried out, the frequency of the procedure, the route of inoculation in any accident, the size of the inoculum, and the prevalence of bloodborne infection in the patient population. For hepatitis B, the immune status of the exposed worker and the promptness and appropriateness of post-exposure immuno-prophylaxis are important variables.

The AIDS epidemic in the United States continues at an unabated rate. In August, 1988 the CDC reported 72,024 AIDS

cases and estimated that 1.0 million to 1.5 million persons
are now infected (16). Figure 1 shows the U.S. incidence of
AIDS by quarter and year of diagnosis reported as of March 31,
1988 from pre-1982 and projected to 1992 from cases reported
as of June 30, 1987 (16).

HIV infection and AIDS continue to show the same demogra-
phics as in the past. The overall cumulative age distribution
per million population (N= 44,795) is shown in Figure 2 (17).

The incidence by sex shows an overall rate for males of
350/1,000,000 and an overall rate for females of 27/1,000,000
(17). The incidence for heterosexual adult and adolescent

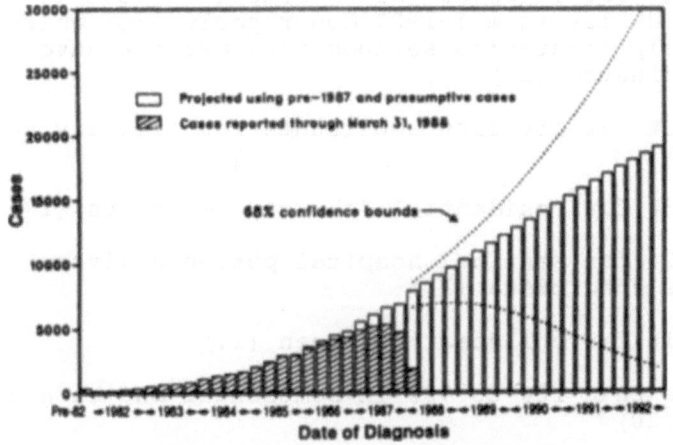

Fig. 1. U.S. incidence of AIDS projected to 1992.
From MMWR 1988;37:551-559 (16).

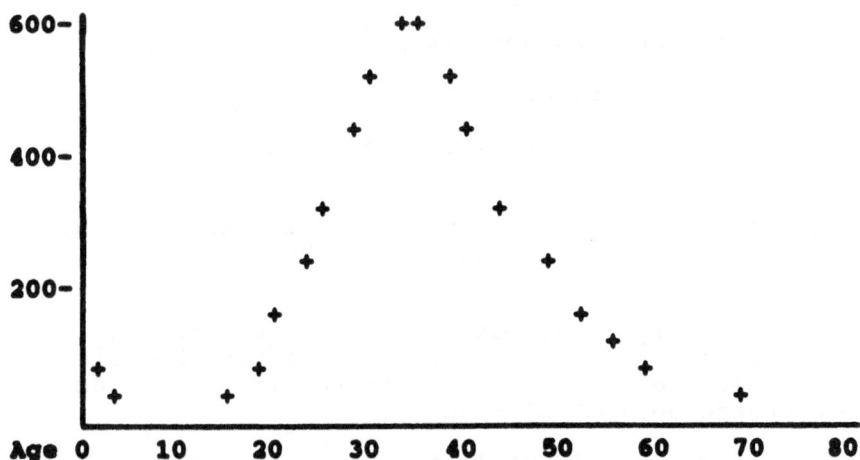

Fig. 2. Cumulative age distribution of AIDS.
Cases per million population. From MMWR
1987;36 (suppl. no. S-6):45 (17).

95

males and females is 73/1,000,000 and 25/1,000,000, respectively (17). The preponderance of males over females is probably due to the relative proportion of homosexual and bisexual men to drug users and others at risk in each group.

The relative incidence by ethnic group shows that blacks have an incidence of 3.0 to 1 over whites and Hispanics have an incidence of 2.6 to 1 over whites (17). This ethnic variability is probably due to the increased risk behavior in these groups.

As a consequence of the variable incidence there is wide variability in the seroprevalence of HIV antibodies in differing patient groups. The prevalence of western blot positive HIV antibodies in the U.S. has been reported to be:

1)   0.04% for voluntary blood donors in 1985-1987 (18) and 0.012% in mid-1987 donor pools from which previously identified seropositive persons have been eliminated (17)

2)   0.15% for civilian applicants for military service (19)

3)   0.33% for residential Job Corps entrants (17)

4)   0.32% for sentinel hospital patients without AIDS-like conditions (17)

5)   0.21% for childbearing women (17)

6)   Less than 10% to 70% in homosexual and bisexual men (17,20)

7)   Less than 5% to 65% in intravenous drug abusers (17,21)

8)   7% to 68% in the long term sexual partners of AIDS patients (17,22)

9)   33% to 92% in hemophiliacs (17,23)

10) 0% to 17% in prisoners (17)

11) 0% to 45% for female prostitutes (17)

12) 0% to 50% for tuberculosis patients (17)

Figure 3 shows the overall incidence of AIDS cases by state, per million population as of November, 1987 (N=44,745) (17).

The geographic incidence of AIDS cases continues to show that population centers with a high rate of risk behavior, male homosexuals and bisexuals, and intra-venous drug abusers have the preponderance of AIDS cases.

The U.S. prevalence of HBsAg in adults is reported to be 0.2% to 1.0% (24), while it is reported to be 35% to 80% and 60% to 80% in homosexual men and intravenous drug abusers, respectively (25).

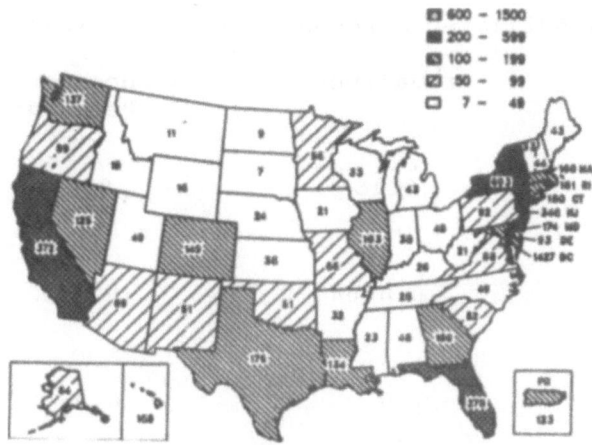

Fig. 3.   Incidence of AIDS cases by state, per million population.   From MMWR 1987;36 (suppl.
no. S-6):45 (17).

The prevalence of HIV infection in the population bears
directly on the risk to health-care workers. The greater the
prevalence, all other factors being equal, the greater the
risk of acquiring a work related infection. One may calculate
the relative and absolute risk of HIV infection after needlestick injury in a direct manner. The seroconversion rate
after needlestick injury has been estimated to be 0.5% (15)
while the rate of accidental needlesticks has been reported to
be 25/100 hospital beds/ year (26). Therefore, the number of
seroconversions to be expected from needlesticks at an institution is:

No. sero-
conversions =   .005 x No. beds (100's) x 25 x Prev. HIV
per year

Table 1 gives estimates of the number of seroconversions
to be expected under different hospital settings.

In April, 1988 the CDC published an update (27) on AIDS
and/or HIV infection among health-care workers with no reported nonoccupational risk factors and for whom case histories
have been published in the scientific literature, from which
Table 2 is condensed.

There have been other less well documented cases reported
and many anecdotal reports. Therefore, the true number of
cases may be larger than accepted by the CDC. Considering the
number of cases of HIV infection and the hundreds of thousands
of HIV bed years of hospital care delivered to HIV infected
patients, the number of occupationally related cases is small.
The apparent low incidence of nosocomial HIV infections coupled with the high variability in geographic prevalence suggests that universal precautions should be modified to suit
the risks present in any single health-care facility. None-

Table 1. Estimated Seroconversions Per Year[*]

Patient Population	Prevalence	Seroconversions Per Year
500 Bed Inner-city Hospital	0.15	0.094
500 Bed Suburban Hospital	0.0004	0.0002
50 Bed AIDS Unit	1.0	0.06

[*]800 HIV bed years are expected to result in one seroconversion due to needlestick injury.

Table 2. Hiv-infected Health-care Workers

Number of Cases	Occupation	Type of Exposure and (Source)
1	Phlebotomist	Muc. Memb. (HIV Pt.)
3	Lab Tech/Wkr	Needle/Muc. (HIV Pt.) Mem./Skin Defect
2	Rsch Lab Wkr	Cut/Skin (Conc Virus)
4	Nurse	Needle/ (HIV Pt.) Muc. Mem. (AIDS Pt.)
2	Home Care	Skin Defect (AIDS Pt.)
9	Not Specified	Needle/Skin (AIDS Pt.) Defect
1	Dentist	Needle (Unknown)

theless, complacency must be avoided since an infection with HIV may be universally fatal (28).

Infection with HBV poses a substantial risk to the health-care worker. The Hepatitis Branch of the CDC estimates that 500-600 health-care workers who are exposed to blood in their job are hospitalized each year, with over 200 deaths per year from early and/or late complications of HBV infection (7). Serologic evidence of past or present HBV infection is seen in 10-40% of health-care workers (29). Therefore, it is strongly urged that all high-risk health-care workers receive HBV vaccine. HBV vaccine should be offered to all such workers by their employers.

AVOIDING THE RISK

The route of accidental exposure is important. Percutaneous exposure is more effective than mucosal or conjunctival exposure in transmitting HIV (14,15). Since it is impossible to know with certainty which patient is infected with HIV or HBV, it is generally recommended that all patients and specified laboratory specimens be treated as if they were known to be infectious for HIV and HBV (1,2).

It has been recommended (1) that HIV antibody screening might be done on all admissions to the hospital or on patients admitted for surgery. Whereas an assay for HBsAg and HBeAg might identify those patients with the highest potential of transmitting HBV (30), the routine admission testing for HIV antibodies is more problematical. HIV antibody assays may not be effective in detecting HIV infected persons in the early latent period (31) and HIV antigen capture assays, HIV genetic probe virus assays, or virus culture are not readily available, costly, and will not necessarily detect all infected individuals (32,33). Therefore, HIV testing on admission may be deceptive, may lower the level of sensitivity to the risk, and be counterproductive (34). All discussions of routine testing are of no consequence to the health-care community since patients must be cared for regardless of their state of infection. Accordingly, it is important that all patients and laboratory specimens be treated as if they are capable of transmitting serious disease.

The Employee's Right-To-Know

Great emphasis has been placed on the patient's right to confidentiality regarding his/her HIV infective status. This is predicated on the fact that public hysteria and lack of information has led to discrimination against HIV infected persons and AIDS patients. This has led to the anachronism of HIV infected patients being admitted to health-care facilities with the knowledge of his/her physician, while the paramedical staff is ignorant of the patients diagnosis. As a corollary, the use of warning labels on patient rooms and specimens has been decried (34) partially because they might compromise the confidentiality of the patients diagnosis. No such paradoxical behavior is associated with HBV infection, even though we now appreciate that HBV is spread by exactly those risk behaviors which are associated with HIV infection.

Under the National Institute for Occupational Safety and Health (NIOSH) and state and local codes employees have a right-to-know of hazards in the workplace. OSHA requires that the employer provide a work-place free of hazards within the scope of the work being performed. Certainly, the presence of an unrecognized HIV infected patient poses a hazard to the employee. Several labor unions representing health-care workers have petitioned OSHA to require employers to offer as hazard free a workplace vis-a-vis HBV and HIV as can be provided. The concept of a hazard free workplace raises the question of which takes precedence; a patients right to confidentiality or a workers right-to-know of hazards in the workplace? While this question is still being debated, it is hoped that the prejudice against persons with AIDS will abate so that the medical community can treat HIV infection without concern for any social stigma or discrimination being attached to the afflicted patient.

MINIMIZING THE RISK - UNIVERSAL PRECAUTIONS

Universal Precautions include practical and familiar procedures to protect the health-care worker and should be followed regardless of any lack of evidence of the patient's infectious status.

Universal precautions as enunciated by the CDC in 1987 and updated in 1988 consist of the following salient points (1,2). (Note: Some portions of this discussion are direct quotes or paraphrases of the CDC recommendations.)

Universal precautions apply to (2):

- (1) Blood

- (2) Tissues

- (3) Body fluids containing visible blood

- (4) Semen

- (5) Vaginal secretions

- (6) The following fluids regardless of visible blood contamination:

    - (a) Cerebrospinal

    - (b) Synovial fluid

    - (c) Pleural fluid

    - (d) Peritoneal fluid

    - (e) Pericardial fluid

    - (f) Amniotic fluid

Universal precautions do not apply to the following unless they contain visible blood (2):

- (1) Feces

- (2) Nasal secretions

- (3) Sputum

- (4) Sweat

- (5) Tears

        (6)   Urine

        (7)   Vomitus

        (8)   Saliva

        (9)   Breast milk

Saliva and breast milk have been reported to contain low concentrations of HBV and HIV in the absence of visible blood contamination (2). Universal precautions are recommended when handling laboratory specimens of saliva, sputum, and breast milk.

Universal precautions should be observed in the dental setting since it can be expected that bleeding will occur during any manipulative procedure. Universal precautions should be practiced when examining mucus membranes of any patient.

Universal precautions are indicated whenever there is an anticipated or actual contact with the body fluids listed above. In addition to the general recommendations, the CDC has published specific recommendations for invasive procedures (including normal delivery), vascular punctures, dentistry, morticians services, laboratory services, autopsies, and dialysis (1,5).

The implementation of universal precautions, which are basically blood precautions, does not affect other types of infection control strategies such as: 1) The identification and handling of infectious waste on the hospital nursing units, the laboratory, the operating rooms, or in an outpatient setting; 2) Protocols for disinfection, sterilization, or decontamination; and 3) Laundry procedures.

Universal Precautions Consist of:

(1)   Barrier protection should be routinely used to prevent skin and mucous membrane contamination with blood, body fluids containing visible blood, or other body fluids to which universal precautions apply.

      The type of barrier protection used should be appropriate for the type of procedure being performed and the type of exposure anticipated.

(2)   Wear gloves when:

    (a)   Touching blood, body fluids, and tissues.

    (b)   Touching all laboratory specimens.

    (c)   Touching mucous membranes and nonintact skin of all patients

    (d)   Handling items contaminated with blood or body fluids.

    (e)   Performing phlebotomy, arterial puncture, skin puncture, and other vascular access procedures.

The following comments concerning phlebotomy are paraphrased from the CDC (2).

The likelihood of hand contamination during phlebotomy depends on several factors: 1) The skill and technique of the worker, 2) The frequency with which the worker performs the procedure, 3) Whether the procedure occurs in a routine or emergency situation, and 4) The prevalence of bloodborne pathogens in the patient population.

The likelihood of infection after superficial skin exposure to blood containing HIV or HBV will depend on: 1) The concentration of virus (viral concentration is much higher for HBV than for HIV), 2) The duration of contact, 3) The presence of skin lesions on the hands of the worker, and 4) For HBV - the immune status of the worker.

Whereas, in universal precautions all blood is assumed to be potentially infective for bloodborne pathogens, in volunteer blood-donation centers the prevalence of infections with HIV and HBV is known to be very low. Some institutions have relaxed recommendations for using gloves by skilled phlebotomists in this setting. Such a policy should be periodically reevaluated. Gloves should be available for workers who wish to use them.

The CDC implies, but does not explicitly state, that in other than voluntary blood-donation centers, where the prevalence of HBV and HIV is known to be very low, that a similar policy on gloves for phlebotomy may be adopted.

The CDC further recommends the following general guidelines:

(a) Use gloves for phlebotomy when the worker has cuts, scratches, or other breaks in his/her skin.

(b) Use gloves in situations where the worker judges that hand contamination may occur, e.g., phlebotomy on an uncooperative patient.

(c) Use gloves for finger and/or heel sticks on infants and children.

The NCCLS recommends that gloves be worn for all skin punctures for the collection of capillary blood since hand contamination is common in this procedure (10).

(d) Use gloves when persons are being trained in phlebotomy.

(3) Change gloves after contact with each patient.

(4) Wear a mask and eye covering, or preferably a face shield, during procedures that are likely to generate droplets of blood or body fluids to prevent exposure of the mucous membranes of the mouth, nose, and eyes.

(5) Wear a gown, apron, or other covering when there is a potential for splashing or spraying blood or body fluids.

(6) Wash hands or other skin surfaces thoroughly and immediately if contaminated with blood, body fluids containing visible blood, or other body fluids to which universal precautions apply.

(7) Wash hands immediately after gloves are removed.

(8) Take extraordinary care to avoid accidental injuries caused by needles, scalpel blades, laboratory instruments, etc. when performing procedures, cleaning instruments, handling sharp instruments, and disposing of used needles.

(9) Place used needles, disposable syringes, skin lances, scalpel blades, and other sharp items into a puncture-resistant biohazard container for disposal. The container should be located as close as possible to the work area.

The NCCLS recommends that phlebotomists should carry puncture-resistant containers with them (10).

(10) To prevent needlestick injuries, needles should not be recapped, purposely bent, cut, broken, removed from disposable syringes, or otherwise manipulated by hand.

Needles used in conjunction with evacuated blood drawing tubes and their adapters pose a special problem. The NCCLS recommends that, alternatively, techniques which will allow one-handed resheathing may be used as shown in Figure 4 (10).

(11) Place large-bore reusable needles (e.g., bone-marrow needles and biopsy needles) and other reusable sharp instruments (sharps) into a puncture-resistant container for transport to the reprocessing area.

(12) Minimize the need for mouth-to-mouth emergency resuscitation procedures. Mouth pieces, resuscitation bags, or other ventilation devices should be used routinely.

13) Health-care workers with exudative lesions or weeping dermatitis should refrain from all patient contact, from handling patient-care equipment, and patient specimens until the condition resolves.

The NCCLS recommends that, alternatively, health-care workers may continue to work if such skin lesions are

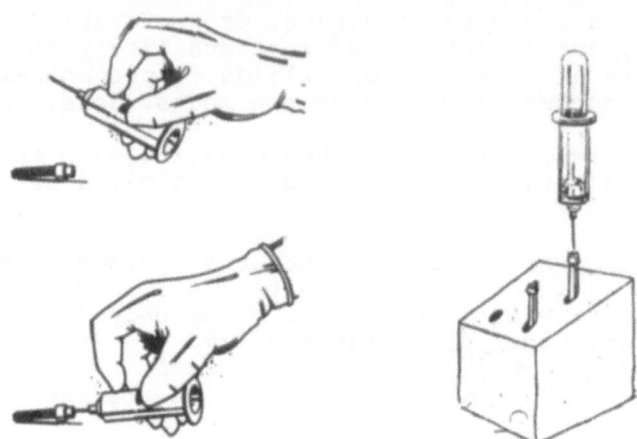

Fig. 4. One-handed techniques for safely resheathing needles on adapters used with evacuated
blood drawing tubes. Left: The needle
sheath may be placed on a stable surface
and "speared" with the needle using one
hand. Right: A retainer may be fashioned
which will hold the needle to allow one-
handed resheathing. Once resheathed, the
needle may be removed and discarded.

covered with a water-proof occlusive dressing to
prevent contamination.

(14) Pregnant women are not known to be at greater risk of
contracting blood-borne infections than other laboratory workers. However, if HBV or HIV infection
develops during pregnancy or if the mother carries
these viruses prior to pregnancy, the infant is at
risk of infection by perinatal transmission. Therefore, pregnant health-care workers should be especially aware of universal precautions.

The CDC makes no special reference to additional risks
posed to immuno-compromised health-care workers. Workers
receiving chemotherapy, adrenalcortical steroid therapy, or
who are biologically immuno-suppressed may be at risk of acquiring one of the opportunistic infections commonly seen in HIV
infected patients and cases of AIDS. The acquisition of a
nosocomial cytomegalovirus infection by a pregnant woman may
endanger the fetus. It may be prudent to limit the patient
contact of such workers.

Gloves

The CDC comments on the selection of gloves are paraphrased
below (2).

Gloves made of thin latex or plastic
are not intended to provide protection from
puncture wounds caused by sharp instruments.
Gloves are intended to cover defects in the

skin of the hands. A single glove will protect the hands from contamination with blood and body fluids. Gloves should be disposable and changed frequently. Gloves should be changed if they become visibly contaminated with blood or body fluids, or if physical damage occurs.

In high risk situations puncture resistance is provided by heavy weight utility gloves such as those used for dishwashing. Stainless steel mesh gloves protect against injury caused by large sharp edges, e.g., knife blades.

The Food and Drug Administration (FDA) has responsibility for regulating the medical glove industry. The CDC recommends the following general guidelines:

1) The material from which the gloves are manufactured should be appropriate for the task being performed. There are no reported differences in barrier effectiveness between intact latex and intact vinyl used to manufacture gloves.

2) Use sterile gloves for procedures involving contact with normally sterile areas of the body.

3) Use examination gloves for procedures involving contact with mucus membranes, unless otherwise indicated, and for other procedures which do not require the use of sterile gloves, e.g., laboratory work.

4) Change gloves between patient contacts.

5) Do not wash or disinfect gloves for reuse. Detergents may cause enhanced penetration of liquids through undetected holes and disinfectants may cause deterioration.

6) Use general-purpose utility gloves (e.g., rubber household gloves) for housekeeping chores involving potential blood contact and for instrument cleaning and decontamination procedures. Utility gloves may be decontaminated and reused but should be discarded if they are peeling, cracked, or discolored, or if they have punctures, tears, or other evidence of deterioration.

The (FDA) is developing standards for gloves. The American Society for Testing and Materials (ASTM) standards permit an acceptable quality limit (AQL) for holes of 2.5 (2.5% defective) for rubber examination gloves and an AQL for holes of 1.5 (1.5% defective) for rubber surgical gloves. The user of gloves should inquire of the manufacturer to ascertain the AQL of the gloves supplied.

The NCCLS recommends double gloving, that is, wearing two pairs of gloves, during autopsies and in other situations where gross contamination of gloves with blood or body fluids is anticipated, such as in the emergency room.

In double gloving, the glove next to the skin may preferably be made of vinyl, which is durable, and the external
glove may preferably be made of latex, which is less likely to
slip.

If the probability of a single latex glove failing is .01,
that is, 1 in 100, then the probability of both gloves failing
simultaneously and in the same place is .0001, or 1 in 10,000.
It is this reduction of the risk of skin contamination that
recommends double gloving in selected circumstances.

When latex gloves are worn for long periods of time or
when they are washed frequently (as during an autopsy) minute
defects, through which blood seeps, commonly develop. Gloves
should be changed at regular intervals during such lengthy
procedures.

BLOOD AND BODY FLUID ISOLATION

Implementing universal precautions eliminates the need for
using the isolation category "Blood and Body Fluid Precautions" previously recommended by the CDC for patients known to
be or suspected of being infected with bloodborne pathogens.
Isolation precautions (e.g., enteric or acid- fast bacillus)
should be used as necessary if associated conditions (e.g.,
infectious diarrhea or tuberculosis) are suspected or diagnosed. However, state or local regulations may require the
use of Blood and Body Fluid Precautions, notwithstanding the
CDC recommendations

BODY SUBSTANCE ISOLATION

Body substance isolation (BSI) is a recently proposed
modification of routine patient care practices augmented by
elements of disease-specific precautions for bloodborne pathogens, most notably HBV and HIV (35). BSI is used for all
patients and does not require a suspicion or diagnosis of a
transmissible disease. For agents which are transmitted by
the airborne route, e.g., tuberculosis, previously described
disease-specific or category-specific isolation procedures
should be used. Although neither the CDC nor OSHA have adopted BSI, the details are given for the readers information.
BSI includes eight components:

1. Gloves

    a) Gloves are worn when there is anticipated con-
        tact with blood, body fluids, secretions, mucus
        membranes, nonintact skin, and all moist sub-
        stances from the body.

    b) Gloves are changed between patients.

    c) When removing gloves and donning new gloves
        between patients, handwashing is not necessary
        if the gloves remain intact.

        If defects are noted in the gloves or if the

hands become visibly soiled then handwashing is needed after removal of the defective gloves.

2. After other types of patient contact, that is with ungloved hands, handwashing should be done before and after each patient contact. Handwashing should consist of 10 seconds of soap and water with friction followed by rinsing with running water.

3. Additional barrier protection (e.g., gowns, face-masks) should be used when blood, body fluids, and secretions are likely to contaminate the skin, mucus-membranes, or clothing of the worker.

4. Soiled reusable articles, linen, infectious waste, and trash should be placed into an appropriate con-tainer to prevent leaking. Single biohazard bags are sufficient, however, if the bag is contaminated on the outside it should be placed into a secondary bag or container.

5. Needles and other sharp instruments should be placed into rigid puncture-resistant containers.

6. Private rooms are indicated for some diseases with airborne transmission, (e.g., pulmonary tuberculo-sis) and for diseases requiring strict isolation.

7. Knowledge of the immune status of the worker is necessary. Vaccinations should be given when neces-sary. Susceptible workers should not enter the rooms of patients with measles, mumps, rubella, and chick-enpox, while immune workers do not need extra barrier precautions for these diseases.

8. There should be a sign stating "Body Substance Isola-tion is for All Patient Care" in every patient room. A "Stop Sign Alert" should be placed on the door of rooms housing patients with airborne diseases in-dicating that disease-specific precautions should be used.

BSI may increase the level of protection in routine pat-ient care while it also relaxes the need for handwashing after removal of intact gloves prior to donning a new pair of glov-es. BSI is gaining wide recognition and is be adopted as an augmentation to universal precautions.

DECONTAMINATION

Decontamination of spills is important in the health-care setting. Decontamination SOPs for spills include:

1. Wear barrier protection.

2. Absorb the spilled material.

3. Clean the site with a detergent solution; absorb the detergent.

4. Soak the site with an appropriate tuberculocidal hospital disinfectant; absorb the disinfectant.

5. Flush the site with water; absorb the water.

6. Dispose of all contaminated materials appropriately.

OCCUPATIONAL EXPOSURE TO HIV AND HBV

It is inevitable that workers will be exposed to HBV and/or HIV. Any parenteral inoculation, mucus-membrane contact, or exposure of non-intact skin poses a substantial risk. Detailed instructions for the post-exposure immunoprophylaxis of HBV infection as recommended by the Immunization Practices Advisory Committee of the CDC have been published (36,37). This prophylaxis includes the selected use of hepatitis B immune globulin and hepatitis B vaccine.

Exposure to HIV infective blood or body fluids causes a lower incidence of infection than does HBV infective blood (12,13,15,28). Nevertheless, the consequences of an HIV infection are substantially worse than those of an HBV infection and workers are justifiably concerned. The following practices are recommended after exposure of health-care workers to blood or body fluids potentially infectious for HIV

In 1988, Burroughs Wellcome Co., the manufacturer of Zidovudine (AZT), instituted a program to determine if prophylactic AZT could decrease the infection rate in health-care workers who are accidentally exposed to blood, blood components, or blood containing body fluids of an HIV infected patient. Two studies in mice suggest that early treatment with AZT may prevent infection with murine leukemia retrovirus and delay the onset and prolong the course of retroviral induced leukemia (38) or neuronal loss due to neurotropic retrovirus (39).

Workers who have been exposed by: 1) needlestick or cut, 2) contamination of non-intact skin, or 3) contamination of mucus membranes are eligible for the study.

The study is a double-blind placebo-controlled trial of a six week course of AZT treatment starting as soon as possible after exposure. The study has received approval from the Human Institutional Review Board (IRB) of the National Institute for Allergy and Infectious Disease and The FDA has issued a waiver of local IRB review. Therefore, individual IRB approval is not needed to participate. It is important not to delay the time of starting the drug after exposure. The drug will be shipped by overnight mail. The participant will be monitored frequently for toxicity for ten weeks and for seroconversion for one year. A complete information packet and a supply of AZT/placebo is available from Burroughs Wellcome Co. by calling 1-800-HIV-STIK.

The CDC recommends (1) that if the exposure occurs from direct contact with a patient, the source patient should be identified and notified of the incident. Voluntary consent should be obtained from the source patient to obtain a blood specimen which should be tested for HIV antibodies as soon as possible. If HIV antibodies are not detected, it may be desirable to test for HIV antigen or for HIV DNA in peripheral blood mononuclear cells using the polymerase chain reaction (PCR) (40). Policies should be developed for testing source patients who cannot consent to being tested (e.g., unconscious patients or minors).

The CDC Agent Summary Statement (41) for HIV recommends that if the exposure occurs from a laboratory contact with a specimen of blood, body fluid, or tissue, the source material (specimen) should be identified and tested for HIV antibodies. If HIV antibodies are not detected, it may be desirable to test for HIV antigen or for HIV DNA by PCR. Prior to testing, all demographic information should be removed from the source material which should be submitted for HIV testing as an anonymous specimen.

If the source patient has AIDS or an HIV infection, is positive for HIV antibodies, HIV antigen, or HIV DNA by PCR, or refuses to be tested for HIV infection; or if the laboratory source material is known to be positive for HIV, tests positive for HIV antibodies, HIV antigen, or HIV DNA by PCR, or is not available for examination:

(1)   The exposed worker should immediately report the exposure to his/her supervisor. The exposed worker should be informed of the AZT prophylaxis protocol and be offered an opportunity to participate.

(2)   The exposed worker should give voluntary consent for a blood specimen to be drawn and tested for HIV antibodies as soon as possible, preferably within 48 hours.

(3)   The worker should be counseled regarding the risk of infection with HIV, and should be evaluated medically for any history, signs, or symptoms consistent with HIV infection.

(4)   The exposed worker should be advised to be alert for the occurrence of any acute febrile illness that develops within 12 weeks of exposure. Such an illness (especially one characterized by fever, rash, or lymphadenopathy or which resembles mononucleosis) may indicate a recent HIV infection.

(5)   Exposed workers who are initially seronegative should be tested for HIV antibodies 6 weeks after exposure. If this test is negative, the worker should be tested at 12 weeks and 6 months after exposure. Most reported seroconversions have occurred between 6 and 12 weeks after exposure.

(6)   The exposed worker should understand how the test results are to be used, the implications of a positive or negative test, and the limits, if any, of confidentiality safeguards regarding the test results.

(7)   HIV may be present in semen, blood, and body fluids before seroconversion. Therefore, during the fol-

low-up period, especially during the first 6 to 12 weeks, the exposed worker should be counseled to follow the recommendations of the CDC and the Surgeon General (42) regarding the prevention of transmission of AIDS, including:

1. Refraining from donating blood or plasma.

2. Informing prospective sex partners of his/her potential exposure to infection so that appropriate precautions can be taken.

3. Avoiding pregnancy during the follow-up period.

4. Informing physicians, dentists, and other health care providers of their potential exposure when seeking medical care so that appropriate precautions can be taken.

5. Cleaning and disinfecting surfaces on which their blood and body fluids have spilled.

6. Not sharing razors, toothbrushes, or other items which could become contaminated with blood.

If the source patient is seronegative, the exposed worker should be tested at 3 and 6 months. However, if the source patient is at high risk of HIV infection, more extended follow-up of the worker may be indicated.

There is a low risk (but not zero) that a seronegative individual may be infectious for HIV. Therefore, if the source patient is at high risk of infection, it may be desirable to repeat testing of the source patient 12 weeks to 6 months after exposure of the worker.

If the source patient cannot be identified, decisions regarding appropriate follow-up should be individualized. The presence of high risk activities in the patient population and the seroprevalence of HIV antibodies in the hospital patients and community should be considered.

Serologic testing for HIV antibodies should be available on a voluntary consenting basis to all health-care workers, particularly those who are concerned that they may have been infected with HIV.

If there is a continuing medical surveillance program for employees, health-care workers may have baseline serum samples collected and stored frozen at $\leq -20\,^\circ C$ for possible future testing. These specimens may prove of epidemiologic value at some future date.

If a patient has a parenteral or mucous membrane exposure to the blood or body fluids of a health-care worker, the patient should be informed of the incident. The same procedures as those recommended above should be followed for the source health-care worker and the exposed patient.

CONCLUSIONS

The decade of the eighties has been the decade of the retrovirus. We have come to know that HIV, the agent of AIDS, is but one of a family of viruses and is currently designated HIV-1. It is to be expected that currently known retroviruses such as human immunodeficiency virus - 2 (HIV-2) and human T cell lymphotrophic virus I (HTLV I) will be joined by many related viruses in being associated with human disease. The decade of the nineties will see a massive increase in the number of AIDS cases as currently HIV-1 infected individuals succumb to the disease. We will also see a wider geographic and host group distribution of AIDS and retrovirus associated disease.

While these prospects are sobering, they need not be depressing. Health-care workers the world over have responded to the AIDS crisis as they have to the plagues of earlier times. With the adoption of prudent safeguards and the exercise of great care in day to day activities, the health-care community will be able to safely cope with the AIDS epidemic.

Universal precautions as we now understand them will be modified to fit the needs of the health-care delivery site, and enforcement should become more rational as the regulatory agencies gain more experience in the field.

REFERENCES

1. CDC. 1987. Recommendations for the prevention of HIV transmission in health-care settings. MMWR. 36 (suppl no. 2S):3s.

2. CDC. 1988. Update: Universal precautions for prevention of transmission of human immunodeficiency virus, hepatitis B virus, and other bloodborne pathogens in health-care settings. MMWR. 37:377-388.

3. Garner, J. S. and B. P. Simmons. 1983. Guideline for isolation precautions in hospitals. Infect. Control 4:245-325.

4. CDC. 1981. Pneumocystis pneumonia-Los Angeles. MMWR. 30:250-252.

5. CDC. 1987. AIDS: Recommendations and guidelines, November 1982 -November 1986. (U.S. Government Printing Office: 1987-732-950/40505), US Department of Health and Human Services, Public Health Service, Atlanta.

6. NCCLS. 1987. Protection of laboratory workers from infectious disease transmitted by blood and tissue; Proposed Guideline. NCCLS Document M29-P. National Committee for Clinical Laboratory Standards, Villanova.

7. Department of Labor, Department of Health and Human Services. 1987. Joint advisory notice: Protection against occupational exposure to hepatitis B virus (HBV) and human immunodeficiency virus (HIV). US Department of Labor, US Department of Health and Human Services, Washington.

8. OSHA. 1988. Enforcement procedures for occupational exposure to hepatitis B virus (HBV), human immunodeficiency virus (HIV), and other blood-borne infectious agents in health care facilities. OSHA Instruction CPL 2-2.44. OSHA, Wash ington .

9. OSHA. 1988. Enforcement procedures for occupational exposure to hepatitis B virus (HBV) and human immunodeficiency virus (HIV). OSHA instruction CPL 2-2.44A, August 15, 1988. OSHA, Washington.

10. NCCLS. 1988. Protection of laboratory workers from infect- ious diseases transmitted by blood, body fluids, and tissue; Tentative Guideline. NCCLS Document M29-T. National Commit- tee for Clinical Laboratory Standards, Villanova.

11. Ziza, J. M., F. Brun-Vezinet, A. Venet, C. H. Rouzioux, J. Traversat, B. Israel-Biet, F. Barre-Sinoussi, J. C. Cher- mann, and P. Godeau. 1985. Lymphadenopathy-associated virus isolated from bronchoalveolar lavage fluid on AIDS- related complex with lymphoid interstitial pneumonitis. N. Engl. J. Med. 313:183.

12. Grady, G. F., V. A. Lee, A. M. Prince, G. L. Gitnick, K. A. Fawaz, G. N. Vyas, M. D. Levitt, J. R. Senior, J. T. Galambos, T. E. Bynum, J. W. Singleton, B. F. Clowdus, K. Akdamar, R. D. Aach, E. I. Winkelman, G. M. Schiff, and T. Hersh. 1978. Hepatitis B immune globulin for accidental exposures among medical personnel: Final report of a multi- center controlled trial. J. Infect. Dis. 138:625-638.

13. Seeff, L. B. 1978. Type B hepatitis after needlestick exposure: prevention with hepatitis B immune globulin. Final report of the Veterans Administration Cooperative Study. Ann. Intern. Med. 88:285-293.

14. Vlahov, D. and B. F. Polk. 1987. Transmission of human immunodeficiency virus within the health care setting. Occupational Medicine: State of the Art Rev. 2:429-450.

15. Gerberding, J. L., C. E. Bryant-LeBlanc, K. Nelson, A. R. Moss, D. Osmund, H. F. Chambers, J. R. Carlson, W. L. Drew, J. A. Levy, and M. A. Sande. 1987. Risk of transmit- ting the human immunodeficiency virus, cytomegalovirus, and hepatitis B virus to health care workers exposed to patients with AIDS and AIDS-related conditions. J. Infect. Dis. 156:1-8.

16. CDC. 1988. Quarterly report to the domestic policy council on the prevalence and rate of spread of HIV and AIDS - United States. MMWR. 37:551-559.

17. CDC. 1987. Human immunodeficiency virus in the United States: A review of current knowledge. MMWR. 36 (suppl. no. s-6)

18. CDC. 1987. Human immunodeficiency virus infection in transfusion recipients and their family members. MMWR. 36:137-140.

19. CDC. 1987. Trends in human immunodeficiency virus infection among civilian applicants for military service - United States, October 1985-December 1986. MMWR. 36:273-274.

20. Lifson, A. R., T. W. Bodecker, J. L. Barnhart, P. M. O'Malley, G. F. Lemp, N. A. Hessol, D. R. Franks, J. J. Cambell, D. Werdegar, H. W. Jaffe, and G. W. Rutherford. 1987. AIDS in the S.F. city cohort study. Program and Abstracts of the 27th Interscience Conference on Antimicrobial Agents and Chemotherapy. American Society for Microbiology, Washington.

21. Robert-Guroff, M., S. H. Weiss, J. A. Giron, A. M. Jennings, I. B. Margolis, W. A. Blattner, and R. C. Gallo. 1986. Prevalence of antibodies to HTLV-I, -II, and -III in travenous drug abusers from an AIDS endemic region. JAMA. 155:151-152.

22. Friedland, G. H. and R. S. Klein. 1987. Transmission of the human immunodeficiency virus. N. Engl. J. Med. 317:1125-1135.

23. CDC. 1987. HIV infection and pregnancies in sexual partners of HIV-seropositive hemophiliac men - United States. MMWR. 36:593-595.

24. Hoofnagle, J. H. 1985. Hepatitis. Principles and Practices of Infectious Diseases, 2nd Ed. G. L. Mandell, R. G. Douglas, and J. E. Bennett (eds.), John Wiley & Sons, New York.

25. CDC. 1984. Adult immunization: Recommendations of the Immunization Practices Advisory Committee (ACIP). MMWR. (suppl no. 1S): 5s-6s.

26. Wormser, G. P, C. S. Rabkin, and C. Jolin. 1988. Frequency of nosocomial transmission of HIV infection among health care workers. N. Engl. J. Med. 319:307-308.

27. CDC. 1988. Update: Acquired immunodeficiency syndrome and human immunodeficiency virus infection among health-care workers. MMWR. 37:229-239.

28. Redfield, R. R. and D. S. Burke. 1988. HIV infection: The clinical picture. Scientific American 259:90-98.

29. Palmer, D. L., M. Barash, R. King, and F. Neil. 1983. Hepatitis among hospital employees. Western J. Med. 138: 519-523.

30. Favero, M. S., J. P. Petersen, W. W. Bond. 1986. Transmission and control of laboratory-acquired hepatitis infection. Laboratory Safety: Principles and Practices. B. M. Miller, D. H. M. Groschel, J. H. Richardson, D. Vesley, J. R. Songer, R. D. Housewright, and W. E. Barkley (eds.), American Society for Microbiology, Washington.

31. Ranki, A., S. L. Valle, M. Krohn, J. Antonen, J. P. Allain, M. Leuther, G. Franchini, and K. Krohn. 1987. Long latency

precedes overt seroconversion in sexually transmitted human immunodeficiency virus infection. Lancet ii:589-593.

32. Casareale, D., S. Dewhurst, J. Sonnabend, F. Sinangil, D. T. Purtilo, and D. H. Volsky. 1984-1985. Prevalence of AIDS-associated retrovirus and antibodies among male homosexuals at risk for AIDS in Greenwich Village. AIDS Res. Hum. Retroviruses 1:407-421.

33. Feorino, P., B. Forrester, C. Schable, D. Warfield, G. Schochetman. 1987. Comparison of antigen assay and reverse transcriptase assay for detecting human immunodeficiency virus in culture. J. Clin. Microbiol. 25:2344-2346.

34. Gerberding, J. L. 1986. The University of California-San Francisco Task Force on Aids: Recommended infection-control policies for patients with human immunodeficiency virus infection. N. Engl. J. Med. 315:1562-1563.

35. Lynch, P., M. N. Jackson, M. J. Cummings, and W. E. Stamm. Rethinking the role of isolation practices in the prevention of nosocomial infections. Ann. Intern. Med. 107:243-250. CDC. 1985. Recommendations of the immunization practices advisory committee. MMWR. 34:313-324,329-335.

36. CDC. 1985. Recommendations of the immunization advisory committee. MMWR. 34:313-324, 329-335.

37. CDC. 1986. Recommendations of the immunization practices advisory committee - Update on hepatitis B prevention. MMWR. 36:353-366.

38. Ruprecht, R. M., L. G. O'Bien, L. D. Rossoni, and S. Nusinoff-Lehrman S. 1986. Suppression of mouse viremia and retroviral disease by 3'-azido-3'-deoxythymidine. Nature 323:467-469.

39. Sharpe, A. H., R. Jaenisch, and R. M. Ruprecht. 1987. Retroviruses and mouse embryos: A rapid model for neuro virulence and transplacental antiviral therapy. Science 236:1671-1674.

40. Ou, C. Y., S. Kwok, S. W. Mitchell, D. H. Mack, J. J. Snisky, J. W. Krebs, P. Feorino, D. Warfield, and G. Schochetman. 1988. DNA amplification for direct detection of HIV-1 in DNA of peripheral blood mononuclear cells. Science 239:295-297.

41. CDC. 1988. 1988 Agent summary statement for human immunodeficiency virus and report on laboratory-acquired infec tion with human immunodeficiency virus. MMWR. 37 (suppl. no. S-4):1S-22S.

42. Koop, C. E. 1987. Surgeon General's Report on acquired immunodeficiency syndrome. U.S. Department of Health and Human Services, Public Health Service, Washington.

SPECIAL PROBLEMS RELATED TO DENTAL OFFICES AND DENTAL SCHOOL CLINICAL FACILITIES

Norman P. Willett

Department of Microbiology and Immunology
Temple University Schools of Medicine
and Dentistry
Philadelphia, Pa

INTRODUCTION

The practice of infection control by the dental profession in both dental offices and clinics has increased dramatically in the past three or four years. Although infection control in dentistry has been practiced to some extent, recent events have greatly accelerated awareness and compliance. These include:

1. The recognition that Hepatitis B is an occupational hazard for dental health care professionals.

2. The real and perceived risks of AIDS.

3. The establishment of guidelines and regulations by various governmental and professional agencies.

Table 1 shows typical goals of infection control in dentistry. They are very similar to those in a hospital setting. A couple have been included which address economic and legal issues. However, a number of unique problems exist for the dental profession in comparison to medicine. Included among these are prolonged close contact with patients, generation of aerosols by the high speed drill, the high incidence of needle and instrument stick, the problems peculiar to sterilization and disinfection of dental instruments, and last but not least, the invisibility of saliva. Crawford dramatically demonstrated saliva spatter by placing red poster paint in the open mouth of a dental manikin and applying a high speed drill. The results are shown in Figs. 1 and 2.

The rationale for infection control techniques in dentistry is based on the Hepatitis B model. Although the main route is from patient to dentist, a number of well documented cases have been caused by dental health care professionals. Of course, the danger also exists of transmission to family if the

Table 1. Goals of Infection Control

1. To reduce the number of available pathogenic microbes to a level where the normal resistance mechanisms of the body may prevent infection.

2. To break the circle of infection and eliminate cross-contamination.

3. To treat every patient as though he or she were infected with a transmissible disease.

4. To protect patients and personnel from infection.

5. To protect income and careers of dental personnel.

6. To protect dental personnel from malpractice suits.

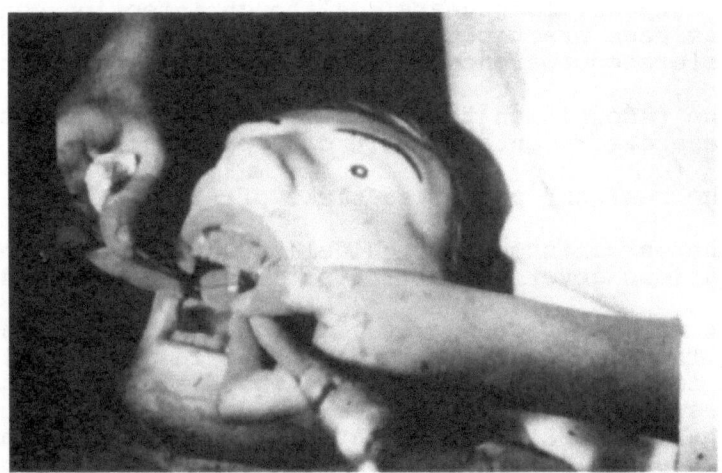

Fig. 1. Simulation of saliva spatter. Courtesy of Dr. James Crawford, University of North Carolina School of Dentistry.

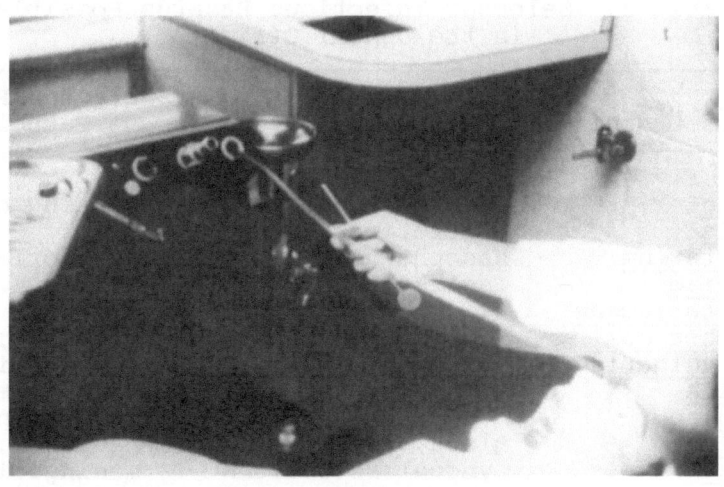

Fig. 2. Courtesy of Dr. James Crawford, Dental Asepsis,
Stoma Press, Seattle, Washington (1979)

dentist contracts the disease. Patient to patient transmission
may also occur, as well as contamination from fomites.

Although Hepatitis B is of prime importance to the dentist,
several other diseases also have an impact on the dental profession. These will be discussed subsequently. Included among
these are HIV infection, Herpes and Legionella. A list of other
infectious diseases which the dentist must also be aware of as
a possibility is shown in Table 2.

Aside from the potential risk of cross-infection, the
dentist has the opportunity to diagnose many systemic infectious diseases which are manifested either initially or early
in their course. Among these are Herpes, mucocutaneous candi-
diasis, Syphilis, Tuberculosis, the acute periodontitis associated with HIV infection, and Hairy Leukoplakia, also associated
with HIV.

Of the four Hepatitis viruses presently known, the number
of cases of Hepatitis B in the U.S. has been steadily increasing since the mid-1960s and now exceeds the number of cases of
Hepatitis A (Fig. 3). The modes of Hepatitis B transmission
can either be percutaneous or nonpercutaneous as shown in
Table 3. Oral surgical procedures, as will be discussed subsequently, are of particular interest in the context of this
presentation. Nonpercutaneous transmission through transfer of
body secretions, e.g., saliva or semen, may not be as efficient
as the percutaneous route; however, this may be compensated for
by continuous exposure.

Table 2. Selected Infectious Hazards Possible
in the Dental Office

Infectious Organism	Transmission	Selected Potential Pathology
Bacteria		
Bordetella pertussis	Nasopharyngeal secretions	Whooping cough
Corynebacterium diphtheriae	Nasopharyngeal secretions	Diphtheria
Enterobacteriaceae		
Escherichia coli Proteus vulgaris Klebsiella pneumoniae	Blood, lesion exudate	Pneumonia, bacteremia, abscesses, wound infections
Haemophilus influenza	Blood, nasopharyngeal secretions	Pneumonia, meningitis, otitis
parainfluenza	"	Conjunctivitis, endocarditis
Mycobacterium tuberculosis	Pharyngeal secretions	Tuberculosis
Mycoplasma pneumoniae	Pharyngeal secretions	Primary atypical pneumonia
Neisseria meningitidis	Blood, nasopharyngeal secretions	Cerebrospinal meningitis
gonorrhoeae	Blood, lesion exudate, nasopharyngeal secretions	Oral lesions, conjunctivitis
Pseudomonas aeruginosa	Lesion exudate	Pneumonia, wound infections
Staphylococcus aureus	Lesion exudate	Suppurative lesions, bacteremia
epidermidis	"	Endocarditis
Streptococcus pyogenes	Blood, nasopharyngeal secretions	Rheumatic, scarlet fever, otitis media, mastoiditis, peritonsillar abscesses, meningitis, pneumonia, acute glomerulonephritis
pneumoniae	"	Pneumonia
viridans group	"	Endocarditis
Treponema pallidum	Exudate from oral lesions	Syphilis
Actinomycosis species (sp) Bacteroides sp Eubacterium sp Fusobacterium sp	Crevicular exudate	Abscesses
Viruses		
Coxsackie virus	Ingestion	Hand/foot/mouth disease, vesicular pharyngitis
Cytomegalovirus	Saliva, blood	Cellular enlargement and degeneration in immunocompromised individuals

Table 2 (cont.)

Infectious Organism	Transmission	Potential Pathology
Epstein-Barr	Saliva, blood	Infectious mononucleosis
Hepatitis		
A	Blood (rare), ingestion	Liver inflammation, jaundice
B	Blood, saliva, tears, semen	Eventual hepatocellular carcinoma in chronic antigen carriers
non-A, non-B	Blood	
delta	Blood	Coinfection with hepatitis B virus (HBV) required
Herpes simplex 1 and 2	Lesion exudate, saliva	Oral lesions, herpetic whitlow, conjunctivitis
Human immuno-deficiency virus (HIV)	Blood	Acquired immune deficiency syndrome (AIDS)
Measles		
rubeola	Nasopharyngeal secretions,	
rubella	blood, saliva, vesicle exudate	Generalized vesicular rash
Mumps virus	Saliva, ingestion	Parotitis, meningitis
Poliovirus	Ingestion	CNS paralysis
Respiratory viruses		
Influenza A and B		Flu,
Parainfluenza	Nasopharyngeal	common cold
Rhinovirus	secretions	
Adenovirus		
Coronavirus		
Varicella	Vesicle exudate	Chicken pox
Fungi		
Candida albicans	Nasopharyngeal secretions	Candidiasis, cutaneous infections
Protozoa		
Pneumocystis carinii	Nasopharyngeal secretions	Interstitial pneumonia in immunocompromised individuals

Adapted from ADA Research Institute, Dept. of Toxicology. J. Am. Dent. Assoc. 117:374 (1988).

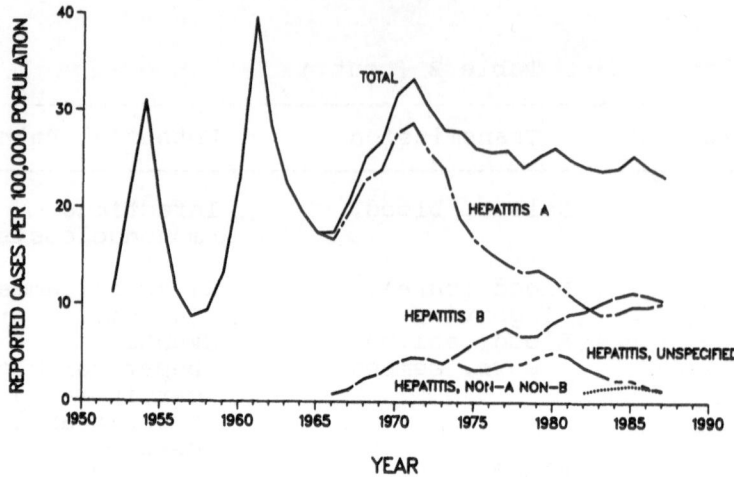

Fig. 3. Summary of Notifiable Disease, 1987, courtesy
of Centers for Disease Control, Atlanta, Ga.

Table 3.   Modes of Transmission of Hepatitis B

Percutaneous Transmission
    Blood and blood products
    Contaminated instruments, or needles
    Tatooing
    Hemidyalysis
    Oral surgery

Non-Percutaneous
    Transfer of body secretions
    Oral (mucosal injury)
    Sexual contact
    Perianal

The higher than average risk of health care professionals
was documented in the early 1950s.[1,2] Mosley and his co-
workers[3] in 1975 evaluated hepatitis as a hazard in general
practitioners. They demonstrated an increased frequency of
infection with Hepatitis B virus among general dentists. In a
further study, Mosley and White[4] showed that specialists with
emphasis in surgical forms of dentistry had a significantly
higher frequency than general dentists. Other investigators[5-7]
have demonstrated that the increased risk for the dental health
care professional starts with patient contact in dental school
and includes not only the dentist but the dental hygienist and
assistants as well (Table 4).

Table 4. Prevalence of Serological Markers of Hepatitis B
Among Dental Personnel[a]

Category	No. positive	Percent positive
Dental hygienists	10/59	16.9
Laboratory technicians	22/155	14.2
Dental assistants	45/350	12.9
Clerical	5/56	8.9
Other	0/9	0
Total	82/629	13.0

[a]Cottone, J. Hepatitis B: transmission and epidemiology in
the dental profession. Proc. Natl. Conf. on Infection Control
in Dentistry, May 1986. USPHS, Atlanta, GA 30333, from data
E. A. Schiff, et al., VA cooperative study on hepatitis and
dentistry (abstr.), Hepatology 2:688 (1982).

Smith and his co-workers[8] compared the risk of Hepatitis
B in dentists, physicians and demonstrated 3.5-fold greater
rate of seropositivity for physicians and dentists than for
volunteer blood donors. They also showed a sharp increase in
prevalence of anti-HB$_S$ with age among physicians and dentists
as compared to a control group of comparable blood donors
(Fig. 4).

Cottone and Gobel[9] in 1983 surveyed the results of a
number of investigators who reported on the prevalence of the
carrier State of U.S. dentists. This summary shown in Table 5
indicated that U.S. dentists have two to six times the normal
risk of developing a Hepatitis B carrier state as the remainder
of the population.

Most of the studies involving the transmission of
Hepatitis B involve patient to dental health professional.[10-12]
However, these have been a number of well documented cases
which involve reports of dentists transmitting Hepatitis B to
patients. Up until 1985 these involved at least 10 outbreaks
involving 172 patients.[13] The highest number of infected
individuals, i.e., 55, attributed to an individual dentist was
reported by Rimland.[14] A 1985 outbreak reported by CDC was
notable in that it was the first outbreak caused by a dentist
which resulted in deaths.[15] The majority of cases involved
oral surgeons, but more critical that all of the outbreaks were
linked to dentists who practiced without routine use of gloves.

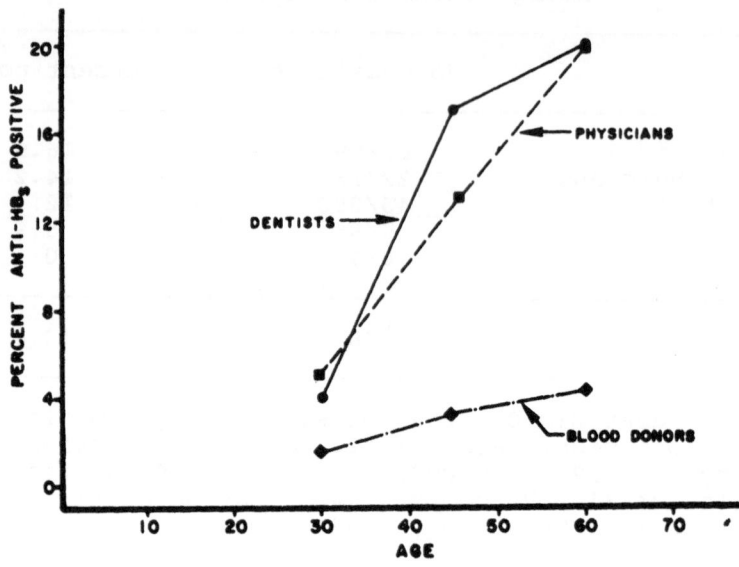

Fig. 4. Prevalence of antibody to Hepatitis B
surface antigen (anti-HB$_S$) in dentists,
physicians and socioeconomically comparable
first time blood donors, J. L. Smith, et al.,
Comparative risk of Hepatitis B among physicians
and dentists, J. Inf. Dis. 133:705-706 (1976).

Table 5.   Prevalence of HBV Carrier State in US Dentists[a]

Authors	No. surveyed	No. HBsAg	Percentage HBsAg[b]
Feldman and Schiff[10]	236	3	1.27
Mosley et al.[3]	1,245	11	0.9
Smith et al.[8]	174	3	1.7
Hollinger et al.[11]	94[c]	4	3.2
Weil et al.[12]	511	4	0.8

[a]J. Cottone, W. Goebel, Hepatitis B: the clinical detection
 of the chronic carrier dental patient and the effects of
 immunization via vaccine, Oral Surg. 56:449-454 (1983).
[b]Approximately 0.3% of US population HBsAg positive.
[c]Dental students.

AIDS

The presence of HIV in saliva has been documented in several studies. Groopman and his co-workers[16] isolated HIV from the saliva of four ARC patients and of health homosexuals. They also observed the virus by electron microscopy in centrifuged saliva of one patient infected with AIDS. Ho and his co-workers[17] concluded that HTLV-III was present infrequently in saliva of AIDS patients. He was only able to isolate the virus from 1% of saliva samples as compared to 56% of blood cultures from 71 seropositive homosexual men. On the other hand, Archibald et al.[18] detected salivary antibodies that reacted with virus-encoded proteins gp 120 and 160 in 70% and 93% of groups of AIDS patients and ARC patients, respectively. Other epidemiologic studies as summarized by Klein[19] reported on almost 4000 dental health care professionals including greater than 8% who treated known AIDS patients. Of these only one dentist tested positive for HIV without any risk other than occupational risk. This later study reported in 1988[20] had a great impact on both the general public and the dental profession due to the banner headlines that appeared in the tabloid press which emphasized the one "doomed dentist" that had acquired HIV infection via occupational transmission.[21] Up to mid-1988 a total of 59 dental professionals had been reported to CDC as being infected with HIV or having AIDS.[21] All of these with the exception of the one dentist previously cited, acquired the disease by nonoccupational means.

The significance of the presence or absence of HIV in saliva still remains undetermined; however, Fultz and her co-workers[23] were unable to infect chimpanzees after oral applications with HIV, whereas intervaginal infection was successful. Fox et al.[24] demonstrated an HIV inhibitory factor in whole human saliva. They showed that the inhibitory activity was secreted by the major salivary glands and did not exist in whole saliva as a result of leakage of serum components through oral lesions or the gingival crevice. They suggested that the inhibitory activity may provide an explanation for the low risk of oral transmission.

Herpes

Herpetic infection of the digits (Herpetic Whitlow) is a recognized hazard of medical and dental personnel. A recent report of an outbreak of HSV in a dental practice in Scranton, Pa., traced the source to a lesion on the fingers of an ungloved dental hygienist.[25] She treated 46 patients when she was infected and 20 of these developed clinical signs and symptoms of severe primary gingival stomatitis which included fever, sore throat and dysphagia.

Legionella

Although Legionella has not generally been considered as a hazard to dental health care personnel, recent reports from dental clinics in the U.S. and Europe have shown increasing antibody titer directly related to the number of years of clinical experience.[26,27] This suggests that Legionella may be present in the dental clinic environment, possibly the water supply, creating an increased risk for clinical personnel and patients.

Barrier Techniques

In view of the real and perceived danger of cross-infection in the dental operatory as documented above, the obligatory use of barrier techniques to prevent these infections becomes self-evident. Of these techniques, none is more important than the use of gloves and masks. As noted previously, the greatest documented risk of transmission of Hepatitis B and Herpes is through direct contact of the patient's mouth with the practitioner's ungloved hand. Masks and glasses serve the function of preventing exposure of the mucous membranes of the practitioner's eyes and face from spattering and splashing by dental procedures, especially the high speed drill.

Other procedures which might properly be classified as barrier techniques, since they prevent cross-infection, are protection against contaminated needles and/or sharp instruments, including their proper handling and disposal, the decontamination of contaminated equipment such as the high speed handpiece, air water syringe and the water lines that service the dental operatory. Normal precautions such as flushing the line with water for a few minutes in the morning are insufficient to reduce counts significantly.[28] Electron micrographs of sections of plastic tubing showed conclusive evidence of biofilms (Fig. 5). Local infections in the oral cavity with Pseudomonas aeruginosa from unit water have been described.[29] Intermittent chlorination has been shown to be an effective method of reducing bacterial counts in dental unit water lines.[30]

Use of protective covering to shield appropriate surfaces in the dental operatory is still another barrier procedure; however, in many cases this is not practical. This necessitates the use of appropriate disinfectants to decontaminate these surfaces.

Liquid sterilants and disinfectants may be divided into four categories: surface disinfectants, immersion disinfectants, immersion sterilants, and hand antimicrobials. Surface disinfectants are usually utilized where the items are too large to be immersed, e.g., cabinets, tables, chairs, etc. Immersion sterilants are utilized for dental instruments of either metal or plastic. These are EPA registered agents that have the capability of killing all living organisms in a recommended immersion time. These contrast with the immersion disinfectants, which are similar to the above only the instruments are disinfected and not washed. Hand or skin antimicrobials are designed to reduce the number of organisms on hands during the specific act of washing or scrubbing.

In considering which liquid disinfectant or sterilant to utilize, it is critical to read and understand the label and/or the accompanying instructions. The efficacy of the agent is perhaps one of the most critical variables and may be interpreted utilizing one or more of several parameters. The shelf life of an agent refers to the time the product remains effective when stored in its original or undiluted form, i.e., on the shelf. In contrast, the use life refers to its effective life after mixing any components or after dilution but without

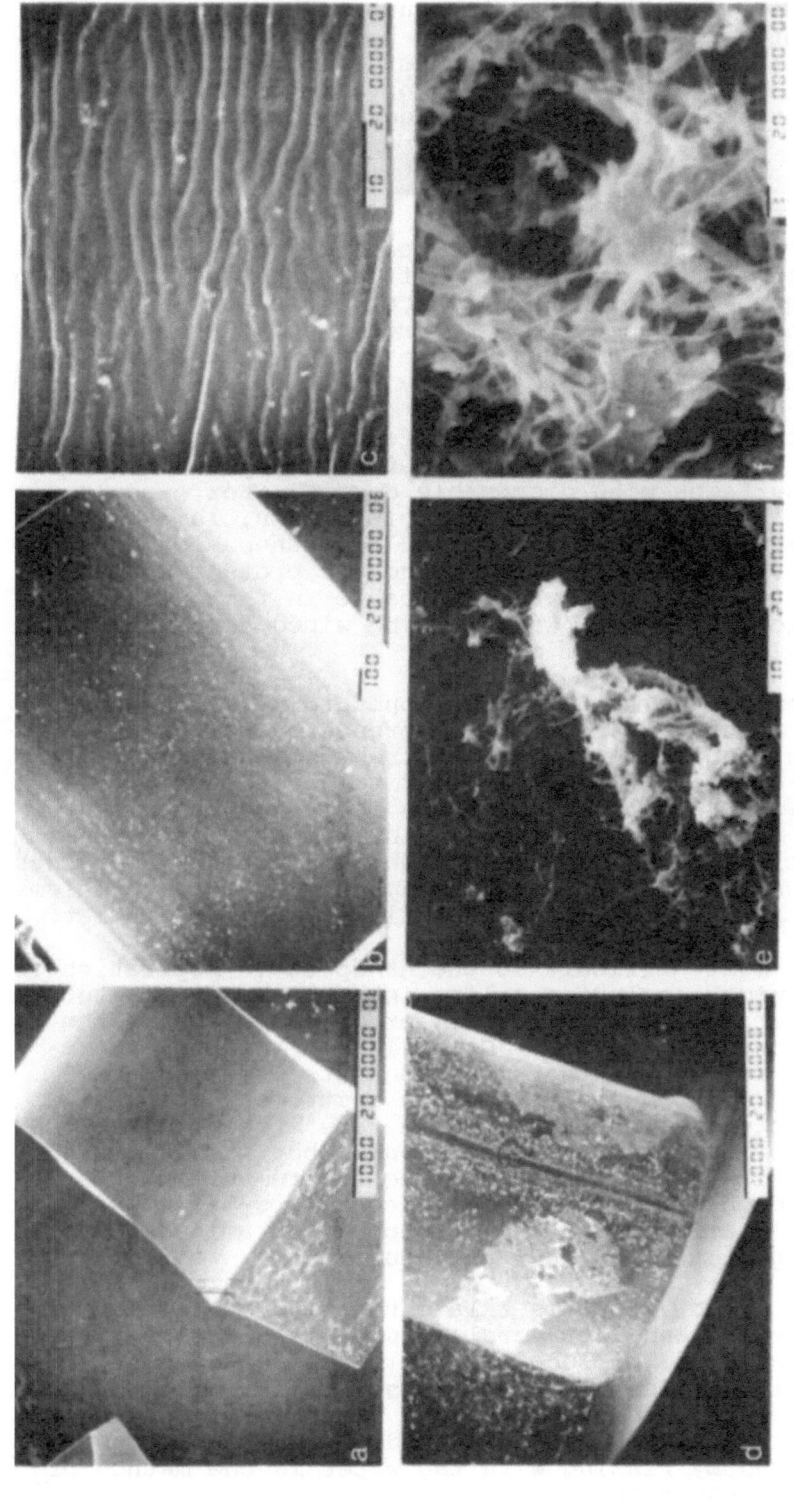

Fig. 5. Biofilms in dental operatory tubing. Courtesy of Dr. J. Mayo, Louisiana State University Medical Center. a. New, unused tubing for air/water syringe, 50x, line = 1,000 μm. b. Used air line from air/water syringe, 75x, line = 100 μm. c. Same as b, 200x, line = 10 μm. d. Used water line from air/water syringe, 50x, line = 1,000 μm. e. Same as d, 2,000x, line = 10 μm. e. Same as d, 7,500x, line = 1 μm.

being utilized. Finally, the re-use life refers to the time that a product is effective in actual use. This is a function of the number of patients treated, the bio-burden (biological material added to the agent) and any water introduced into the agent. In addition, if the EPA registered label does not specify any of these variables the product must be discarded daily.

Operatory Asepsis - Special Problems

The many surfaces of the dental operatory make the prevention of contamination an extremely difficult and time consuming procedure. Figure 6, depicting a typical dental operatory, gives an idea of its complexity. Figures 1 and 2 shown previously demonstrated the magnitude of the potential contribution by aerosol spatter which is produced by the high speed drill.

Although there are a number of effective surface disinfectants (usually an iodophor), the most effective procedure is to cover as many surfaces with disposable covers. For example, an instrument tray may be covered with a bib napkin; lamp handles are usually wrapped with foil (borrowed from those for wrapping baked potatoes). A common problem is how to protect patient records which are usually nearby on a table in the operatory. A simple plastic book cover, which can be wiped down with an appropriate disinfectant will suffice.

The problem of the multi-surface operatory is further complicated by instrumentation, which presents challenging decontamination problems. Prior to current emphasis on infection control techniques, many dentists resisted autoclaving instruments on a number of grounds, including economic, unfamiliarity, rusting and dulling of instruments, etc. Most of these obstacles have been overcome due to current guidelines of the ADA, OSHA and CDC; however, a number of special problems of decontamination and sterilization still exist for dentists.

One of the most difficult is the handpiece. Aspiration of contaminated fluids, including saliva into the handpiece tubing is a common occurrence. Installation of check valves in the water coolant can prevent this. Sterilizable handpieces are available. The problem is that many dentists are concerned that autoclaving will ruin an expensive instrument. Industry has also not gotten its message through very effectively either to the schools or the private practitioner that sterilizable handpieces do exist. In some cases, the instrument may be sterilizable and the practitioner may negate its effectiveness by a post-sterilization lubricant (probably contaminated with a lubricant-loving organisms, such as Pseudomonas).

A second group of problems relate to the disinfection of impression materials and intra-oral prosthesis. These impression materials are made of such materials as alginates, hydrocolloids and polysulfides, which obviously cannot be autoclaved and may be susceptible to attack by the usual chemical disinfectant. Brief chair-side sprays of iodophors or sodium hypochlorite, followed by rinsing with tap water is one method of disinfection in these cases.

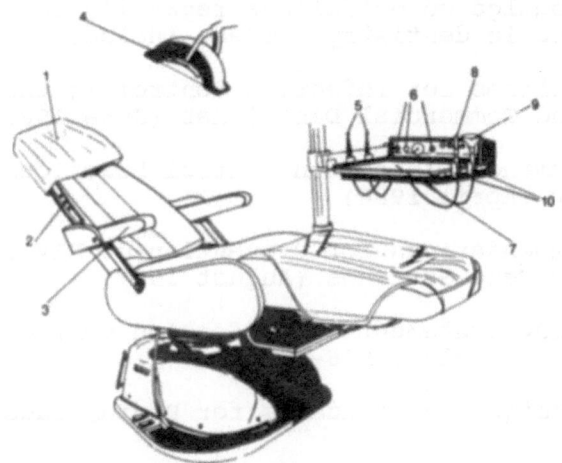

```
             Order of Preference
  (S = sterilization, C = covering, D = disinfection)
     1   Head rest cover          C
     2   Chair switches           C, D
     3   Arm rest                 C, D
     4   Lamp handles             S, C, D
     5   Handpieces               S, C, D
     6   Controls                 C, D
     7   Table surface            C, D
     8   Evacuator tip            S or dispose
         Evacuator grip           C, D
     9   Cuspidor funnel          C, D
    10   Holders                  C, D
```

Fig. 6. Surface components of an operatory and their
        aseptic care. James Crawford, "Clinical Asepsis
        in Dentistry," Oral Medical Press, Mesquite,
        Tex. (1986).

The dental technician who usually works in a lab geographically removed from the dentist must be as fully aware of the possibility of cross-infection as any dental health care personnel in direct patient contact. He must follow the same stringent precautions when handling and formulating intra-oral appliances as the dentist.

Plastic materials, if not disposable, can be adequately sterilized by such agents as acid or alkaline glutaraldehyde, provided the appropriate exposure time is used. Usually this is 6-10 hours in order to achieve sterilizing conditions, a factor which many dentists choose to ignore either because of lack of knowledge or economic reasons.

Impact of Guidelines

Since 1982, a number of governmental and professional

agencies have promulgated guidelines regarding infection control procedures in dentistry. These include:

1. ADA Guidelines for Infection Control in the Dental Office and Commercial Dental Lab (June 1982)

2. CDC Recommended Infection Control Practices for Dentistry (April 1986)

3. CDC Recommendations for Prevention of HIV Transmission in Health Care Settings (August 1987)

4. AADS policy statements on infection control practices in dental schools (1986-1988)

5. ADA Accreditation Standards for Dental Education Program

6. OSHA Standards (1988)

As you will note, most of these pronouncements have had only the force of recommendations and guidelines. Last year, OSHA stepped into the picture and it appears that most of these previous recommendations will now have the full force of a governmental regulatory agency behind them. The need for enforcement even after all of the guidelines, educational programs and the fear factor generated by HIV, is reinforced by a number of compliance studies. Studies in the San Francisco area before and 4 months after the report by Klein of the occupationally HIV infected dentist showed significantly greater compliance with recommended infection control guidelines.[31] A later study in July of 1987 showed even more striking increases; however, even with these substantial increases in compliance, only 73% reported using gloves with all patients and fewer than 60% wore masks.[32]

An earlier study in late 1986 of dentists in the Los Angeles area demonstrated that although most dentists are aware of the high risk groups in the population, less than one third of those who attended Continuing Education courses could not name any oral signs and symptoms of AIDS.[33]

These studies, together with a number of other earlier compliance studies demonstrate that dental health care workers have certainly increased dramatically their compliance of infection control procedures due to a massive educational effort by professional and governmental bodies. However, as these recent studies demonstrate, there is still a significant lack of compliance and apathy that exists.

It remains to be seen what the impact of the OSHA regulations on the compliance will be. Enforcement of institutional compliance will be difficult in its own right, but eventually will be similar to enforcement in hospitals. It remains to be seen what impact these regulations can possibly have on the office practitioner, although some scattered reports have come through relating to this.

REFERENCES

1. M. L. Trumbell and D. J. Greiner, Homologous serum hepatitis: an occupational hazard to medical personnel, J. Am. Med. Assoc. 145:965-967 (1951).
2. E. B. Byne, Viral hepatitis. An occupational hazard of medical personnel, J. Am. Med. Assoc. 195:362-364 (1966).
3. J. W. Mosley, V. Edwards, G. Casey, A. Redeker and E. White, Hepatitis B infection in dentists, N. Eng. J. Med. 243:729-734 (1975).
4. J. W. Mosley and E. White, Viral hepatitis as an occupational hazard of dentists, J. Am. Dent. Assoc. 90:992-997 (1975).
5. W. M. Goebel and G. L. Getnich, Hepatitis B virus infection in dental students: a two year evaluation, J. Oral Med. 34:33-36 (1979).
6. S. P. James and R. E. Samplener, Hepatitis B in the dental setting: dental hygienists, J. Md. State Dent. Assoc. 21:26 (1978).
7. E. R. Schiff, et al., VA cooperative study on hepatitis in dentistry, (Abstract) Hepatology 3:688 (1982).
8. J. L. Smith, J. E. Maynard, K. R. Baguist, I. L. Doto, H. M. Webster and M. J. Sheller, Comparative risk of hepatitis B among physicians and dentists, J. Inf. Dis. 133:705-706 (1976).
9. J. A. Cottone and W. Goebel, Hepatitis B: the clinical detection of the chronic carrier dental patient and the effects of immunization via vaccine, Oral Surg. 56:449-454 (1983).
10. R. E. Feldman and E. R. Schiff, Hepatitis in dental professionals, J. Am. Med. Assoc. 232:1228-1230 (1972).
11. F. B. Hollinger, et al., Hepatitis B prevalence within a dental student population, J. Am. Dent. Assoc. 94:521-527 (1977).
12. R. B. Weil, et al., A hepatitis serosurvey of New York dentists, N.Y. State Dent. J. 43:587-590 (1977).
13. J. Cottone, Hepatitis B treatment and epidemiology in the dental profession, Proc. Natl. Conf. Infection Control, Centers for Disease Control, October 1986.
14. D. Rimland, et al., Hepatitis B outbreak attributed to an oral surgeon, New Eng. J. Med. 296:953-958 (1977).
15. F. E. Shaw, C. L. Pearl, P. J. Coleman, S. C. Hadler and J. E. Maynard, Lethal outbreak of Hepatitis B in a dental practice, J. Am. Med. Assoc. 255:3260-3264 (1984).
16. J. Groopman, S. Z. Salahudin, M. G. Sarngadharan, P. Markham, M. Gonda, A. Sliska and R. C. Gallo, HTLVIII in saliva of AIDS related complex and healthy homosexual men at risk for AIDS, Science 226:447-449 (1984).
17. D. Ho, R. Byington, R. Schooley, T. Flynn, T. Rota and M. S. Hirsch, Infrequency of isolation of HTLVIII virus from saliva in AIDS, N. Eng. J. Med. 313:1606 (1985).
18. D. W. Archibald, L. Zon, M. F. Groopman and M. F. Essex, Antibodies to human T-lymphotropic (virus type III) in

saliva of acquired immunodeficiency syndrome (AIDS) patients and persons in risk for AIDS, _Blood_ 67:831-834 (1986).

19. R. S. Klein, _in_: "Occupational Risk of HIV, Perspectives on Oral Manifestation of AIDS," P. Robertson and J. Greenspan, ed., p. 12-27, PSG Publishing Co., Littleton, Mass. (1988).

20. R. S. Klein, J. Philan, K. Freeman, K. Schable, G. Friedland, N. Trieger and N. Steigbegel, Low occupational risk of human immunodeficiency virus infection among dental professionals, _N. Eng. J. Med._ 318:86-90 (1988).

21. New York Daily News, Friday, June 5 1987.

22. Personal communication, Centers for Disease Control, Atlanta, Ga.

23. P. Fultz, H. McClure, H. Dougharty, A. Brodie, C. McGrath, B. Swenson and D. Francis, Vaginal transmission of human immunodeficiency virus (HIV) to a chimpanzee, _J. Inf. Dis._ 154:896-900 (1986).

24. P. Fox, G. Wolff, C. K. Yek, J. Atkinson and B. Bann, Saliva inhibits HIV-1 infectivity, _J. Am. Dent. Assoc._ 116:635-637 (1988).

25. J. P. Manzella, J. H. McConville, M. A. Mengus, E. M. Swirkosz and M. Arens, An outbreak of herpes simplex virus type 1 gingivostomatis in a dental hygiene practice, _J. Am. Med. Assoc._ 252:2019-2022 (1984).

26. P. G. Fotos, H. N. Westfall, I. S. Snyder, R. W. Miller and B. M. Mutchler, Prevalance of Legionella specific IgG and IgM antibody in a dental clinic population, _J. Dent. Res._ 64:1352-1355 (1985).

27. F. F. Reinthaler, F. F. Mascher and D. Stanza, Serologic examination for antibodies against Legionella species in dental personnel, _J. Dent. Res._ 67:942-943 (1988).

28. J. A. Mayo, K. M. Oertleny and S. C. Anececo, Bacterial biofilm: contamination, source in air-water syringes, _J. Dent. Res._ 68(special issue):284 (1989).

29. M. V. Marlin, Significance of bacterial contaminants of dental unit water systems, _Brit. Dent. J._ 163:152 (1987).

30. N. W. Feihn and K. Hemikson, Methods of disinfection of the water system of dental units by water chlorination, _J. Dent. Res._ 67:1499-1508 (1988).

31. B. Gerbert, _in_: Dentistry and AIDS: An Educational Challenge, Perspectives on Oral Manifestations of AIDS, P. Robertson and J. Greenspan, ed., p. 185-198, PSG Publishing Co., Littleton, Mass. (1988).

32. B. Gerbert, V. Badner, B. Maguire, et al., Perceived personal risk: impact on dentists' infection control behavior, _Abst. J. Dent. Res._ (special issue) 67:256 (1988).

33. K. Atchison, T. Dolan and H. Meetz, Have dentists assimilated information about AIDS, _J. Dent. Ed._ 51:668-672 (1987).

OCCUPATIONALLY ACQUIRED HUMAN IMMUNODEFICIENCY VIRUS-1 INFECTION

IN HEALTH CARE WORKERS: A REVIEW

Robyn R.M. Gershon and David Vlahov

Johns Hopkins University School of Hygiene and Public Health
Department of Environmental Health Sciences
615 North Wolfe Street, Baltimore, MD   21205

INTRODUCTION

The Public Health Service has estimated that the current number of individuals infected with human immunodeficiency virus, type 1 (HIV-1) in the United States is 945,000 to 1,410,000 people (1). Within one to five years 20-30% of those infected will develop acquired immuno-deficiency syndrome (AIDS); recently it has been postulated that perhaps as many as 99% of infected individuals will eventually develop AIDS (2). Over time, it is inevitable that increasing numbers of HIV-1 positive individuals will enter the health care system, and this in turn will result in an increase in the potential for exposure to HIV-1 among health care workers.

Speculation and uncertainty about the actual risk of infection to health care workers still exists. Even though epidemiological studies have shown that HIV-1 is not readily transmitted in the occupational setting, well publicized anecdotal reports have tended to obscure the true risk of infection. Some evidence is available that health care workers may be quite fearful of the risk of acquiring HIV-1 in the workplace. A recent survey by Link and co-workers showed that 25% of medical interns in their study population would refuse to care for AIDS patients if they could (3). A study of University of California, San Francisco housestaff attitutes towards AIDS found that 80% of those surveyed were at least "mildly anxious" about caring for AIDS patients (44). Over 80% of the respondents indicated that fear of acquiring the illness was paramount to their concerns. A recent study of laboratory workers who handle HIV-1 infectious material found workers to have high levels of both fear and concern regarding their occupational risk of infection (R. Gershon, B. Curbow, D. Vlahov, in preparation).

The purpose of this paper is threefold: to review data that address the issue of risk of occupationally acquired HIV-1 infection; to outline risk management strategies to minimize the risk; and to discuss policy issues regarding health care workers who may have been exposed to HIV-1 in the workplace.

131

RISK ASSESSMENT

Assessment of risk for HIV-1 infection among health care workers involves several components. The components include identification of types of exposure, estimating risk of exposure, and finally, estimating risk of infection given an exposure. The types of exposures involving HIV-1 among health care workers most frequently include: 1) direct contact with an open wound, 2) direct contact with mucous membranes and 3) parenteral inoculation. There is no evidence for HIV-1 transmission via aerosols, and therefore exposure to such aerosols is not relevant. Of the three routes, parenteral inoculation, most often occurring through needlestick injuries, has been well studied in health care workers. McCormick and Maki have shown that needlestick injuries account for a large number of health care worker accidents (5). The majority of needlestick accidents result from recapping needles, the improper disposal of needles and syringes and mishaps associated with venipuncture and the administration of injections (6).

Needlestick injuries are a serious source of exposure to HIV-1 infectious material among health care workers. In 1986, McCray showed that 68% of the health care workers enrolled in a nationwide prospective surveillance program were exposed to HIV-1 through needlestick injuries (7). In an updated analysis of the surveillance data, the Centers for Disease Control (CDC) reported that approximately 89% of the HIV-1 exposures were caused by percutaneous exposure to infected blood (8). CDC noted that 7% of the exposures resulted from contamination of open wounds, and 5% from contamination of mucous membranes. It should be noted, that the CDC "Needlestick Cooperative Study" intermittently selected for percutaneous injuries between 1983 when the project was initiated and 1987 when the enrollment criteria was expanded. Even so, other data, including a sample of health care workers not selected for route of exposure, found needlestick injuries to be the leading cause of HIV-1 exposure. More recently, Ramsey reported on a small group of exposed workers, 70% of whom were exposed through the percutaneous route (10). Gerberding and co-workers also found that many workers had reported multiple needlestick injuries involving HIV-1 infected blood; one such worker sustained a total of 11 accidental needlestick injuries involving patients with AIDS (9). Henderson and co-workers reported similar findings (12).

Given the high frequency of occupational exposures, particularly by percutaneous routes, a major concern is estimating the risk of infection given an exposure. Several epidemiological studies designed to assess the risk of infection have estimated the risk to be less than 1% (Table 1). The most extensive collection of data on this subject is provided by the nationwide CDC "Needlestick Cooperative Study". Their most recent data show four seroconversions in 963 exposed health care workers meeting the criteria for occupational exposure (8,31,43)*.

The National Cancer Institute (NCI) study of laboratory workers currently has over 225 exposed laboratory workers enrolled, two of whom are seropositive (11). The well designed prospective study at the Clinical Center of the National Institutes of Health (NIH) recently reported on an exposed worker who seroconverted out of over 800 exposed workers enrolled in that study (12,13). Recently, Gerberding and

---

*Four seroconversions were noted among workers who had sustained percutaneous exposures. No seroconversions were found in 103 workers with exposures resulting from either contamination of mucus membranes or non-intact skin.

Table 1. Seriological Studies on Health Care Workers Exposed to HIV-1

A. *Prospective Studies

Study Author	Total # Enrolled in Study	Total # With Parenteral/ Mucosal Exposure	Total # With Sero- conversion
Geddes (England/ Wales, 1986) (21)	89	89	0
CDC (USA, 1986, 1987) (8,43)	1613	963	4
Henderson, (NIH, 1986) (12,13,43)	800+	332	1
Kuhls, et al (Los Angeles, 1987) (19)	292	25	0
McEnvoy (Canada, 1982) (20)	150	150	0
Gerberding, et al. (San Francisco, 1987,1988) (9, 14)	270	129	1
Weiss, et al. (NCI, 1988) (11)	265	10	1
Ramsey, et al. (Texas, 1988) (10)	44	44	1
Totals	3305	1742	8

Rate of Seroconcersion = 8/1742 = .0046
4.6 cases per thousand workers with either parenteral or mucosal
exposure.

---

*Subjects with at least 2 serololgical evaluations, one negative at
baseline, and the other performed at leat 180 days after exposure.

**Table 1. Seriological Studies on Health Care Workers Exposed to HIV-1
(continued)**

B. Cross-Sectional Studies

Study Author	Total # Enrolled in Study	Total # With Parenteral/ Mucosal Exposure	[o]Total # With Sero- conversion
Jones, et al (England, 1985) (46)	21	NS	-
Hirsch, et al (Boston, 1985) (16)	85	33	-
Weiss, et al (MD, NY, 1985) (17)	361	42	xx-
Mann, et al (Zaire, 1986) (23)	2492	NS	^-
Gerberding, et al (San Francisco, 1986) (18) (Dentists)	264	22	-
Klein, et al (NYC, 1988) (22) (Dentists)	1309	1230	#-

Totals

o = Since only a single serologic specimen was tested, no
seroconversions are noted.
xx = Three subjects reported seropositive (no documented
seroconversions).
^ = 152 workers were found to be seropositive, however, there were no
documented seroconversions and no occupational factors were
significant in the regression analysis.
# = One subject reported seropositive (no documented seroconversion).
NS = Not stated

co-workers reported one seroconversion among more than 235 exposed workers in their San Francisco study (14).

Nursing personnel have been identified in a number of studies as having the greatest potential for exposure (63% of enrolled workers in the CDC study, 21% in Gerberding study), followed by physicians, medical students, laboratory technicians and phlebotomists (33,9).

The total number of health care workers meeting the minimum requirements for enrollment (i.e. negative baseline, known exposure and posititve post-serology from all reported cross-sectional and prospective studies is now in excess of 1500 (Table 1). From these studies, 8 documented seroconversions in health care workers have been identified (Table 2). These 8 cases have met the criteria for an occupationally acquired HIV-1 infection as defined by Vlahov and Polk (15). The criteria include:

1. Occupational exposure to HIV-1 materials
2. No other established risk factors present
3. Negative pre-exposure HIV-1 serology
4. Positive post-exposure HIV-1 serology

In addition to these ten reports of seroconversion among survey subjects, there are another ten fully documented anecdotally reported seroconversions in health care workers. These anecdotal case reports are summarized in Table 3 and are not included in the estimation of risk of infection among exposed workers because these individuals were not enrolled in a surveillance survey at the time of their exposure; therefore, the denominator for these additional case reports is unknown.

Similarly, an additional five anecdotal reports are also not included in the estimation of risk (Table 4). These case reports probably represent occupational transmission, yet since they lack adequate documentation they are therefore considered "apparent transmissions". One such "apparent transmission" is described in a report by Weiss and co-workers concerning a research technologist working in an HIV-1 production laboratory (11). The worker was found to be seropositive upon entry into the study. However, evidence is strong that this case truly represents occupationally acquired infection since the virus cultured from the seropositive worker was found to share an RNA genotype pattern that was _virtually identical_ to the one used in the employee's laboratory (11). Because of this compelling evidence this case is included in the estimate of risk.

In summation, the risk of exposure is substantial, especially for medical/nursing personnel; this risk is related to the HIV-1 seroprevalance rate of the patient population within a particular health care setting. Percutaneous exposure to HIV-1 infected blood, especially involving needlestick injuries, is most likely to lead to transmission. Infection, though rare, is clearly documented under these circumstances.

RISK MANAGEMENT STRATEGIES

Regulatory Controls

Until very recently regulations affecting the occupational health and safety of health care workers were either non-existent or too broadly defined to be useful. However, in August 1987, the Occupational Safety and Health Administrations (OSHA) issued a proposed set of

Table 2. **Summary of Occupationally Acquired HIV-1
Infections in Health Care Workers
(as of November, 1988)***

Year	HCW sex/ Occupation	Country/ Source	Circumstances of Exposure

Documented Seroconversions from Epidemiological Studies

1. 1986	Female HCW	USA/Stricof et al. (25)	During an emergency procedure a deep intramuscular injury with a large bore needle was inflicted by a co-worker. The needle was contaminated with blood from an AIDS patient.
2. 1987	NS	USA/CDC (8)	Deep needlestick injury, from a large bore needle also inflicted by an co-worker during a resusitation attempt on an AIDS patient.
3. 1987	NS	USA/CDC (8)	Two needlestick injuries, AIDS patient and HIV-1 infected patient sources.
4. 1987	NS	USA/ Gerberding, Henderson. (14)	Needlestick injury involving blood from an AIDS patient.
5. 1988	HCW	USA/ Ramsey (10)	Needlestick injury with material from an HIV-1 infected patient.
6. 1988	laboratory research technician	USA/ Weiss (11)	Laboratory worker in an HIV-1 production laboratory with a history of non-intact skin on the hands. There were episodes of HIV-1 contamination in the work area.
7. 1988	HCW	USA/CDC (31)	While attempting to fill a glass blood tube, there was an accidental self-injection of several milliliters blood from a patient with AIDS.
8. 1988	Male technician	USA/NIH (13)	A tube of blood from a patient with AIDS broke in his hand while he attempted to recap the tube.

Year	HCW sex/ Occupation	Country/ Source	Circumstances of Exposure

Documented Anecdotal Case Reports[*]

1.	1984	Female nurse	England/ anonymous (24)	Needlestick injury while recapping a needle used to draw blood from an AIDS patient.
2.	1986	Female nurse	France/ Oksenhendler et al. (26)	Superficial needlestick scrape while recapping a needle that was contaminated with bloody pleural fluid from a patient infected with HIV-1.
3.	1986	Female student nurse	Martinque/ Neisson- Vernant et al (27)	Needlestick injury during blood drawing from an AIDS patient.
4.	1986	Female Health Provider	USA/CDC (28)	Home health care provided to young son with multiple, severe medical problems. The child was HIV-1 positive following multiple transfusions. The mother (a trained health provider) had numerous exposures to the child's blood and body fluids. The mother did not wear gloves and frequently did not wash her hands after they were contaminated.
5.	1987	Female HCW	USA/CDC (29)	Held bloody gauze for 20 minutes on bleeding site during an emergency procedure. The patient had AIDS.
6.	1987	Female phelbotomist	USA/CDC (29)	While filling a 10ml vacuum blood tube with blood from an HIV-1 infected patient, the top flew off and blood splattered the worker on the mouth and face. The worker had on gloves and eyeglasses. Facial acne was present.

Table 3. Summary of Occupationally Acquired HIV-1 Infections in Health Care Workers (as of November, 1988)*

Year	HCW sex/ Occupation	Country/ Source	Circumstances of Exposure
7. 1987	Female Medical Technologist	USA/CDC (29)	While manipulating an apheresis machine, a large blood spill occurred. The worker was not wearing gloves. Dermatitis was present on one ear, and this may have been touched during the clean up procedures.
8. 1988	laboratory research worker	USA/ Weiss (11)	Laboratory worker in an HIV-1 production facility suffered a cut to the hand with a blunt stainless steel cannula. The cannula had been used to clean out a centrifuge rotor and it was contaminated with concentrated HIV-1 material.
9. 1988	female nurse	Italy/ Gioannini (30)	Mucous membrane exposure with material from an HIV-1 infected patient.
10. 1988	Male Navy corpsman	USA/ Wallace (41)	Punctured a fingertip while disposing of a phlebotomy needle that had been used on an HIV-1 positive patient.

Table 4. **Summary of Occupationally Acquired HIV-1 Infections in Health Care Workers (as of November, 1988)***

Year	HCW sex/ Occupation	Country/ Source	Circumstances of Exposure

Anecdotal Cases Without Documented Seroconversion

	Year	HCW sex/ Occupation	Country/ Source	Circumstances of Exposure
1.	1985	Home Health Care Provider	England/ Grint (33)	^ Home health provider with dry, cracked skin and numerous exposures to AIDS patient's blood and body fluids.
2.	1985	Female HCW	USA/CDC (17,34)	^ History of two accidental needlesticks with blood from AIDS patients. At the time of enrollment into surveillance study, this worker was HIV-1 positive.
3.	1985	Male HCW	USA/CDC (17,34)	^ History of accidental needlestick injury while processing blood from a multiply transfused leukemic patient. History of a needlestick injury while processing pooled platelets. At the time of enrollment into surveillance study, this worker was HIV-1 positive.
4.	1988	Male Dentist	USA/ Klein (23)	^ Numerous exposures,cuts. No gloves worn.
5.	1988	Male laboratory technician	Mexico/ Ponce de Leon (32)	^ Deep cut with processed + blood from numerous donors many of whom were HIV-1 positive.

---

^ = these workers do not have fully documented occupationally acquired HIV-1 infection since baseline sera was not available for testing. Nevertheless, these cases probably do represent occupational infection.
+ = This worker died of AIDS.
HCW = Health care worker.
* = Only those case histories that have been published in the scientific literature have been cited.

regulations that address the specific issue of occupationally acquired blood-borne infections in the health care system. The impetus for this new ruling is a direct result of employee pressure to have mandated strict compliance with the CDC guidelines on blood borne infectious agents in the workplace. The proposed standard, entitled "Protection Against Occupational Exposure to Hepatitis B Virus and Human Immunodefiency Virus " addresses a wide area to ensure the protection of the health care worker (35). Key aspects of the proposal include the provision of adequate personal protective equipment and supplies, the development of educational and training programs, the development of a medical surveillance program for employees, and the maintenance of all records pertinent to the overall health and safety program.

The standard is currently being enforced through the OSHA General Duty Clause. The standard provides for the inspection of facilities as well as the setting of fines for those health care institutions that are not in compliance. Recently, the directive has been revised so that the hospitals' own infection control program and its implementaiton is used as a basis for citing or fining. If the infection control program is lacking then CDC guidelines will be used as a standard. A permanent standard is anticipated by the end of 1988.

Procedural Controls

Recommended procedures should be adopted as standard practice to minimize the risk of occupationally acquired HIV-1. A very brief description of prevention measures are listed:

1. "Universal" precautions should be followed when handling all patients and patient specimens (36). CDC has recommended that "Universal" precautions do not apply to feces, nasal secretions, sputum, sweat, tears and vomitus unless these are visibly contaminated with blood. While epidemiological studies have not indicated that certain body fluids such as saliva, sweat, urine and vomitus can transmit HIV-1 or hepatitis B virus, many health care workers have found it prudent to take adequate precautions when handling these materials (42). Universal precautions are intended to supplement standard infection control procedures and practices and must therefore be used with either category or disease specific systems (37). Universal precautions include the use of appropriate gloves whenever contact with blood or blood tinged body fluid is likely to occur, irrespective of knowledge of the patient's serostatus. Hands should be washed after patient contact and after removing gloves.

2. In accordance with recent EPA recommendations, all articles contaminated with blood or body fluids, irrespective of knowledge of serostatus, should be properly decontaminated before they are terminally disposed or reprocessed.

3. All blood spills should be cleaned up promptly using an effective disinfectant such as a 1:100 dilution of household bleach (.5% sodium hypochorite).

4. Protective clothing (gloves, gowns, aprons, sleeves, masks, shields, eyewear) should be worn where appropriate to the level of anticipated exposure.

5. A "sharps" program must be in effect for each health care

setting, including home health care. Needles, including butterfly needles must __never__ be recapped, resheathed or disarmed. "Sharps" must be discarded into a sturdy "sharps" container and then terminally disposed according to local and state regulations.

6. All incidents, accidents and near accidents ideally should be immediately reported to supervisory and medical staff.

Administrative Controls

A key risk management strategy is the development and implementation of effective occupational health and safety policies and procedures.

Management of Exposed and Infected Health Care Workers

Routine screening of health care workers for antibodies to HIV-1 is not recommended. Although the Elisa Immunoassay (EIA), has very high specificity and sensitivity, a low prevalence of infection in a population such as health care workers will produce a low positive predictive value. False positive tests will result in the additional worry and expense of confirmatory testing.

Recommendations have been made with respect to the proper follow - up procedures in the event of an HIV-1 exposure in a health care worker (8,39). Serum for baseline HIV-1 serological testing should be drawn at the time of exposure. Written permission from the patient source must be obtained prior to testing the patient's blood for HIV-1. If the patient is found to be HIV-1 seropositive, or if the patient refuses to be tested, then the exposed employee should be retested for antibody to HIV-1 on a periodic basis (six weeks, three months, six months and twelve months). These employees should be offered both medical and psychological counseling and support. Exposed health care workers must have complete assurance that an investigation of their incident will be handled strictly confidentially and that medical records will be coded to ensure confidentiality. Only one medical staff person should have knowledge of HIV-1 exposures. In some instances supervisors may be aware of the incident and this may be especially important in helping to prevent future accidents. In addition, health care employers may request the enrollment of the exposed worker in one of the many ongoing prospective studies. In all cases the number of individuals aware of the identify of the exposed worker should be severely limited.

In the event that exposure leads to infection, it is an absolute requirement that this information be kept strictly confidential. Sanctions against such breaks in confidentiality must be in place and well publicized in the workplace. At this point both workers compensation and OSHA must be involved and various strategies have been suggested for the maintenance of confidentiality. In general, this is best handled by advanced planning and forethought. HIV-1 positive employees and their families should be provided with complete and thorough medical and psychological counseling. This should include information on health benefits, workers compensation benefits, life insurance, and the availability of ongoing psychological support.

In general, workers infected with HIV-1 need not be restricted from active patient care (39). Invasive procedures may require some modification (i.e. double-gloving) or in some instances curtailment of certain procedures such as the use of wire sutures during surgery. Health care workers with communicable diseases should naturally refrain from direct patient care. Similarly, infected workers with weeping or exudative lesions should not have direct contact with patients.

Pregnant workers have raised special concerns about working with AIDS patients or their specimens. If the proper precautions are followed, the pregnant health care worker should not be at any additional risk of exposure.

Management of HIV-1 Infected Patients

Testing all patients' blood for HIV-1 has been proposed as a means of minimizing the risk to health care personnel. There are a number of objections to this approach. In many health care situations (emergencies, comatose patients, etc.), it may not be feasible to obtain the informed consent of the patient. There is also the problem of obtaining test results rapidly so that they are of use to emergency room personnel; clearly, even rapid test results are of no value to out-of-hospital (offsite) emergency workers. In addition to serious concerns regarding confidentiality of patients, the financial burden of mass patient testing needs to be considered (i.e. who will pay for the testing?). Finally, the possibility of a false negative test result (e.g., the patient is infected but has not yet produced antibodies) may in fact increase the risk of exposure because of a false sense of security among hospital personnel.

Many workers have expressed concern over some of the other infectious diseases commonly found in AIDS patients. Gerberding, and co-workers, have shown that health care workers are not at increased risk of acquiring cytomegalovirus or HBV from exposure to patients with AIDS (9).

CONCLUSION

There is ample evidence of frequent exposure to HIV-1 among health care workers. A number of epidemiological studies have shown that while the exposure rate may be high for this population, the rate of transmission is low. To date, eight documented seroconversions have been detected out of more than 1500 health care workers enrolled in studies. Many of the exposures and subsequent infections were preventable. An analysis of the exposures in the CDC cohort shows that approximately 40% of these could have been prevented by following standard recommended procedures (7).

Prevention strategies should include the use of environmental, administrative, and procedural controls. OSHA has recently implemented regulatory controls to encourage the strict implementation of standard recommended procedures. Although the content of these recommendations are similar to those advised by the United States Public Health Service, the OSHA Standard has the force of governmental authority to implement and monitor infection control programs. The proposed OSHA Standard places responsibility for compliance with the employer. The employee also has a responsibility under the General Duty Clause to comply with all health and safety regulations. Prior to this standard, health care professionals were considered responsible for themselves. Under the new proposed standard, employers can institute stepwise punitive actions culminating in job termination for those employees who consistently refuse to follow the recommended safety precautions. This type of relationship has not existed before.

At the present time, prevention and management strategies remain our best defense against HIV-1 infection in the health care workplace. Strict adherence to standard procedures and recommended precautions is essential in the effort to minimize risk. Training should be accomplished with a full appreciation and utilization of behavioral modification and persuasive communication techniques. The professional expertise of health psychologists should be called upon as necessary to assist in the development of truly effective training interventions. Standard policies should be adopted and in place to effectively handle critical issues such as the management of exposed and infected employees. Infection control practitioners have long provided considerable leadership in the arena of risk reduction among health care workers in terms of policy development and implementation. The OSHA Standard provides a renewed stimulus to health care administrators to call upon and support infection control practitioners in their efforts to conduct effective control programs.

Acknowledgements

The authors wish to thank Dr. Eric Sansone for his review of the manuscript, and to Dr. Al Saah and Dr. Tim Townsend for their helpful comments and suggestions.

REFERENCES

1. J. W. Curran, H. W. Jaffe and, A. Hardy, Epidemiology of HIV infection and AIDS in the United States, Science 239:610-616 (1988).

2. D. I. MacDonald, Public Health Service plan for the prevention and control of acquired immunodeficiency syndrome (AIDS) and the AIDS virus, Pub Hlth Rep. 101:341-348 (1986).

3. R. N. Link, A. R. Feingold, and M. H. Charap, Concerns of medical and pediatric house officers about acquiring AIDS from their patients, AJPH. 78:455-459 (1988).

4. J. L. Dienstag and D. M. Ryan, Occupational exposure to Hepatitis B virus in hospital personnel: Infection or immunization?, Am. J. Epidemiol. 115:26-39 (1982).

5. R. McCormick and D. Maki, Epidemiology of needlestick injuries in hospital personnel, Am. J. Med. 70:928-932 (1981).

6. D. L. Palmer, M. Barash, R. King and F. Neir, Hepatitis among hospital employees, West J. Med. 138:519-23 (1983).

7. E. McCray, The cooperative needlestick surveillance group: Occupational risk of the acquired immunodefficiency virus syndrome among health care workers, N. Engl. J. Med. 314:1127-32 (1986).

8. CDC, Recommendations for prevention of HIV transmission in health - care settings. MMWR 36 (Suppl. 2S) (1987).

9. J. L. Gerberding, C. E. Bryant-Leblanc, and K. Nelson, Risk of transmitting the human immunodeficiency virus to health care workers exposed to patients with AIDS and AIDS related conditions, J. Infect. Dis. 156:1-8 (1987).

10. K. M. Ramsey, E. N. Smith, and J. A. Reinarz, Prospective evaluation of 44 health care workers exposed to human immunodeficiency virus -1, with one seroconversion, <u>Clin. Res</u>. 36:22A (1988).

11. S. H. Weiss, J. J. Goedert, and S. Gartner, Risk of human immunodeficiency virus (HIV-1) infection in laboratory workers, <u>Science</u> 239:68-71 (1988).

12. D. K. Henderson, A. J. Saah, and B. J. Zak, Risk of nosocomial infection with human T-cell lymphotrophic virus type III/ Lymphandenopathy associated virus in a large cohort of intensively exposed health care workers, <u>Ann. Int. Med</u>. 104:644-647 (1986).

13. D. Barnes, Health workers and AIDS: Questions persist, <u>Science</u> 241:161-162 (1988).

14. J. L. Gerberding and D. K. Henderson, Design of rational infection control policies for human immunodeficiency virus infection, <u>J. Infect. Dis</u>. 156:861-864 (1987).

15. D. Vlahov and B. F. Polk, Transmission of human immunodeficiency virus within the health care setting, Occupational Medicine State of the Art Reviews, 2:451-470 (1987).

16. M. S. Hirsch, G. P. Wormser, and R. T. Schooley, Risk of nosocomial infection with human t-cell lymphotropic virus (HTLV-III), <u>N. Engl. J. Med</u>. 312:1-4 (1985).

17. S. H. Weiss, W. C. Saxinger, and D. Rechtman, HTLV-III infection among health care workers: association with needle-stick injuries, <u>JAMA</u> 254:2089-93 (1985).

18. J. L. Gerberding, C. E. Bryant-Leblanc, and D. Greenspan, Risk to dentists from exposure to patients infected with AIDS virus (abstract). In: Abstracts of the 26th Interscience Conference on Antimicrobial Agents and Chemotherapy, Washington, DC. ASM 283 (1986).

19. T. L. Kuhls, S. Viken, and N. Parris, Occupational risk of HIV, HBV and HSV-2, <u>AJPH</u> 77:1306-1309 (1987).

20. M. McEnvoy, K. Porter, and P. Mortimer, Prospective study of clinical, laboratory and ancillary staff with accidental exposures to blood or other body fluids from patients infected with HIV, <u>Br. Med. J</u>. 294:1595-1597 (1987).

21. A. Geddes, Risk of AIDS to health care workers, <u>Br. Med. J</u>. 292:711-712 (1986).

22. R. S. Klein, J. Phelan, and K. Freeman, Low occupational risk of human immunodeficiency virus infection among dental professionals. <u>N. Engl. J. Med</u>. 318:86-90 (1988).

23. J. M. Mann, H. Francis, and T. C. Quinn, HIV seroprevalence among hospital workers in Kinshasa, Zaire, <u>JAMA</u>, 256:3099-3102 (1986).

24. Anonymous, Needlestick transmission of HTLV-III from a patient infected in Africa, Lancet, 2:1376-1377 (1984).

25. R. L. Stricof and D. L. Morse, HTLV-III/LAV seroconversion following a deep intramuscular needlestick injury, N. Engl. J. Med. 314:1115 (1986).

26. E. Oskenhendler, M. Harzic, and J. M. Le Roux, HIV infection with seroconversion after a superficial needlestick injury to the finger, N. Engl. J. Med. 315:582 (1986).

27. C. Neisson-Vernant, S. Arfi, and D. Mathez, Needlestick HIV seroconversion in a nurse, Lancet 2:814 (1986).

28. CDC, Apparent transmission of HTLV-III/LAV from a child to a mother providing health care, MMWR. 35:76-79 (1986).

29. CDC, Update: Human immunodeficiency virus infection in health care workers exposed to blood of infected patients, MMWR. 36:285-289 (1987).

30. P. Gioannini, A. Sinicco, and G. Cariti, HIV infection acquired by a nurse, Eur. J. Epidemiol. 4:119-120 (1988).

31. CDC, Update: Acquired immunodeficiency viorus infection among health care workers, MMWR, 37:229-239 (1988).

32. S. Ponce de Leon, G. Sanchez-Mejorda, and M. Zaid-Jacobson, AIDS in a blood bank technician in Mexico City, Infect. Control Hosp. Epidemiol. 9:101-102 (1988).

33. P. Grint and M. McEvoy, Two associated cases of AIDS, Comm. Dis. Report. 42:4 (1985).

34. CDC, Update: Evaluation of human T-lymphotropic virus type III/Lymphandenopathy associated virus infection in health care personnel, United States, MMWR, 34:575-578 (1985).

35. Joint Advisory Notice, HBV/HIV, Federal Register, Pages 41818-41824 October 30, 1987.

36. CDC, Update: Universal precautions for prevention of transmission of human immunodeficiency virus, hepatitis B and other bloodborne pathogens in health care settings, MMWR, 37:377-388 (1988).

37. J. S. Garner and M. S. Favero, Guidelines for handwashing and hospital environmental control, Atlanta U.S. Department of Health and Human Services, Public Health Service, CDC HHS Publication No. 99-117 (1985).

38. CDC/NIH, Biosafety in Microbiological and Biomedical Laboratories, 2nd edition, HHS Publication #84-8395 (1988).

39. J. E.Conte, Infection with HIV in the hospital, Epidemiology, infection control, and biosafety considerations, Ann Intern Med. 105:730-736 (1986).

40. T. L. Kuhls and J. D. Cherry, The management of health care workers' accidental parenteral exposure to biological specimens of HIV seropositive individuals, Infection Control. 8:211-213 (1987).

41. M. R. Wallace and W.O. Harrison, HIV seroconversion with progressive disease in health care worker after needelstick injury, Lancet. ii:1454 (1988).

42. A. R. Lifson, Do alternate modes for transmission of human immunodeficiency virus exist? A review, JAMA. 259:1353-1356 (1988).

43. R. R. Marcus, Surveillance of health care worker exposed to blood from patients infected with the human immunodeficiency virus, N. Eng. J. Med. 319:1118-1123 (1988).

44. M. Cooke, Housestaff attitudes toward the acquired immunodeficiency syndrome, AIDS and Public Pol. 3:59-60 (1988).

45. P. Jones, P. Hamilton, HTLV-III antibodies in haematology staff, Lancet, 1:216 (1985).

REAL AND PERCEIVED RISKS OF INFECTION TO HEALTH CARE WORKERS:
WILL UNIVERSAL PRECAUTIONS WORK?
PANEL DISCUSSION

Kathleen Meehan Arias

Infection Control Section
Frankford Hospital
Philadelphia, Pennsylvania

INTRODUCTION TO THE PANEL MEMBERS

The members of the panel were Stanley Bauer, M.D. (Bronx-Lebanon
Hospital Center, Bronx, NY), Robyn Gershon, M.H.S. (Johns Hopkins
University, Baltimore, MD), Martin S. Favero, Ph.D. (Centers for Disease
Control, Atlanta, GA), Bruce Kleger, Dr.P.H. (Pennsylvania Department of
Health Laboratories, Lionville, PA), and Norman Willett, Ph.D. (Temple
University Schools of Medicine and Dentistry, Philadelphia, PA). The
moderator was Kathleen Meehan Arias, M.S. (Frankford Hospital,
Philadelphia, PA).

INTRODUCTION TO THE DISCUSSION

This symposium, "Infection Control: Dilemmas and Practical Solutions,"
consisted of fifteen sessions plus this panel discussion. During the
sessions nine of the speakers used the term "universal precautions." Most
of you would probably agree that universal precautions is now one of the
most demanding issues facing clinical and laboratory scientists who are
working in the areas of infectious diseases and infection control. And,
based on comments and questions from the speakers and the audience at this
conference, it certainly has become a controversial topic.

Why the controversy? For hundreds of years man has known that blood
and body fluids can transmit disease and has used various types of barrier
precautions such as protective garments, quarantine, and isolation to
prevent disease transmission from an infected person to others. Many of
you have seen sketches of the garb worn by physicians to protect themselves
during the Great Plague. We all know that gowns, gloves, masks, and
isolation precautions are widely used and accepted in hospitals for the
prevention of disease transmission. So why is there now an increased
emphasis on the fact that all blood and body fluids are potentially
infectious? This is not a new idea. However, concern about exposure to
AIDS, to human immunodeficiency virus (HIV), is new. The AIDS epidemic has
greatly increased both health care workers' and the general public's
awareness of infections and infection control. Before AIDS most health
care workers had a laissez-faire attitude about disease transmission in the
workplace. But now many of us know workers who have gone to the other
extreme—some of whom are so fearful of infection that they have left the
health care field. Many of us in infection control and infectious diseases

are perplexed by the reactions of our co-workers and colleagues to the universal precautions issue. Robyn Gershon's study, presented this morning, shows that we need to know more about health care workers' knowledge, beliefs, and attitudes before we can develop educational interventions to effectively change behaviors.

Before we begin the panel discussion, I would like to give an overview of how the concept we call "universal precautions" has evolved over the past few years. Studies show that the occupational risk of acquiring HIV infection in the health care setting is very low (less than 1%) following percutaneous or mucous membrane exposure to HIV-infected blood or body fluid. Fewer than twenty cases of documented occupationally-acquired HIV infection have been reported in the literature.[1] Despite the apparently low risk, there is presently no antiviral treatment or vaccine for HIV. Therefore, guidelines were developed to protect health care workers from infection.

Early guidelines published by the Centers for Disease Control (CDC) and others [2,3,4], outlined precautions that should be taken when handling blood and body fluids from persons judged likely to have AIDS. As the magnitude of asymptomatic infection with HIV became recognized, policy makers developed guidelines [5,6,7,8], which advised health care workers that the blood and body fluids of all patients are potentially infectious. Most of these guidelines stressed protection against HIV but hardly mentioned hepatitis B virus (HBV). Concern over HIV transmission frequently overshadowed the fact that the danger of occupational exposure to unrecognized hepatitis B virus carriers has been well-recognized since the late 1970's and greatly affected many health care workers' acceptance of the HBV vaccine. It wasn't until 1986 that many of the guidelines for protection against HIV exposure stressed the importance of routinely handling all blood and body fluids carefully. The term "universal precautions" became widely known in mid-1987 when several groups including the CDC [9] and the American Hospital Association [10] used it in published guidelines for reducing occupational exposure to HIV and HBV. According to the CDC [9] implementation of universal precautions eliminates the need for using the isolation category of "Blood and Body Fluid Precautions"; however, it does not eliminate the need for other isolation precautions such as enteric precautions for patients with infectious diarrhea. The CDC did not intend for universal precautions to replace widely-accepted isolation precaution techniques.

In 1987 a system called body substance isolation or BSI was described.[11] BSI was proposed as an alternative system to the CDC isolation precautions [12] commonly used in hospitals to prevent nosocomial infections. Because BSI and "universal precautions" were widely introduced around the same time, many people have confused the two. Both are concepts used to prevent the transmission of infectious organisms in the health care setting; however, the methods for implementing the two do differ. Whereas BSI was proposed to replace disease-specific or category-specific isolation practices, universal precautions, as intended by the CDC, should be used in addition to isolation precautions.

Although it appears quite simple and sensible, the concept of universal precautions is a substantial departure from routine, established practices in health care facilities and has created much confusion and indecision among health care workers, health care employers, educators, and policy makers.

In October of 1987 the U.S. Department of Labor and the U.S. Department of Health and Human Services issued a joint advisory notice [13] on protection against occupational exposure to HBV and HIV. This was

followed shortly thereafter by the Occupational Safety and Health
Administration (OSHA) advance notice [14] that it was preparing to develop
regulations that would require employers to implement programs aimed at
reducing occupational exposure to HBV and HIV. Health care facilities with
effective Infection Control Programs already had procedures in place to
prevent the spread of HBV, HIV, and other organisms in the workplace well
before the CDC and the Department of Labor issued their guidelines in
1987. However health care organizations became pressured to implement
something called universal precautions when OSHA's Compliance Safety and
Health Officers were instructed to use the term when inspecting health care
facilities [15]. In June of this year the CDC issued an update [16] to clarify
its previous recommendations for universal precautions and in August OSHA
revised its enforcement procedures for occupational exposure to HBV and
HIV. [17]As we heard this morning the National Committee for Clinical
Laboratory Standards has recently released its guidelines titled
"Protection of Laboratory Workers from Infectious Disease Transmitted by
Blood and Tissue." However, there is presently no consistent agreement on
how, where, and by whom universal precautions should be implemented. Until
there is a consensus on the who, what, where and why issues, "universal
precautions" will probably continue to be a controversial topic.

PANEL DISCUSSION

MODERATOR: It's now time for the "universal precautions brouhaha" as one
of our panel participants called it yesterday. There are a few questions
that we hope to answer and a few that we hope to raise. One of the
questions that we hope to answer is "What is universal precautions?" Many
of us have different views of just what universal precautions is. Dr.
Bauer made a good point this morning when he said that universal
precautions is a practice. It really is a practice and we've all
interpreted it in slightly different ways. The panel topic is "Real and
Perceived Risks of Infection to Health Care Workers: Will Universal
Precautions Work?" Another question we wanted to address is "What are the
real risks to health care workers?" Robyn Gershon addressed that question
well this morning. A third question is "What are the perceived risks to
health care workers?" This is your chance to let us know what you've found
at your facilities. And lastly, "Will universal precautions work?"...or is
it just "bureaucratic effluvia" as someone suggested yesterday?

MODERATOR QUESTION: When Dr. Bauer showed his slides this morning, I think
that a lot of you noticed that under the Body Substance Isolation system
you do not have to wash your hands after removing gloves following patient
care. Would one of the panel members like to comment on this?

ANSWER (Robyn Gershon): Dr. Favero and I usually disagree but I am firm on
this. I think you should wash your hands after everything. If it takes a
little extra time and it means that you have to wear hand cream so be it.

ANSWER (Dr. Favero): I would agree with that.

ANSWER (Dr. Kleger): I sat in on some of the NCCLS discussions, and as Dr.
Bauer said this morning we had a wide range of input. We had hospital
people, government people, pathologists, medical technologists, and
infection control people and this was a point of much discussion. I
believe the first version of the NCCLS document said thatyou could wash
your hands with your gloves on rather than changing gloves. There was some
pressure to change that and it has been changed. Some people felt that you
would get tears in your gloves. I don't think the final word is in. We
have to deal with the real world and the real world says that people don't
like to wash their hands between patients, especially if they've been
wearing gloves, because they don't think it's necessary. If they're going

to put on another pair of gloves before they see another patient they feel they don't have to wash their hands. Somebody will come up with a study that shows one way or another what counts, but right now we're dealing with too much anecdotal information.

COMMENT (Dr. Kleger): Before we take questions from the audience, I'd like to make a comment. We talk about the relative risks of HIV in different parts of the country, San Francisco, New York and Miami being high and Fargo, North Dakota being low. Aside from HIV don't we have to worry about anything else? An 89 year old woman has some probability of having HBV infection. So is AIDS the only thing we have to be concerned about? When we wrote the NCCLS documents, to justify [universal precautions] on all levels we had to take all blood-borne, tissue-borne, and fluid-borne infections into consideration. I think if you just look at the perceived risks of HIV infection in your patient population you're overlooking everything else. HBV infection is age-related so the older the person the greater the likelihood. An 89 year old woman may have had seventeen blood transfusions during her life. So I think this certainly has to be taken into consideration in your levels of protection needed.

QUESTION (Dr. Kleger): I would like to ask Robyn a question. Since the beginning of HIV infection many black people thought that they could not get infected and that it was a white man's disease. Were there any ethnic or racial differences that you saw in your study [of risk taking beliefs and behaviors of health care workers handling HIV-1]?

ANSWER (R. Gershon): We didn't ask that demographic question. It was an internal policy decision. I wish we had asked it.

COMMENT (Dr. Willet): About two years ago at an ASM meeting, Jim Curran gave a talk on AIDS and on risks and relative risks. He compared the relative risk of getting AIDS from a blood transfusion to the risks that we take in everyday life such as driving an automobile or walking down the street. He noted that the risk of getting HIV from a blood transfusion was less than the risk of getting killed in an automobile accident on the way to the hospital to get that transfusion. The whole idea of relative risk impressed me.

COMMENT (R. Gershon): The risk of [HIV] infection was also less than the risk of dying from the surgery itself. Of interest in the risk literature: for most people the "magic number" is a fatality that results from one out of a thousand times of whatever it is you're doing--for instance, sky diving. If it results in a fatality of one out of one thousand times, that risk is unacceptable to most people. We're asking health care workers to accept a risk of what, four per thousand. I think that's interesting.

COMMENT (Dr. Favero): That's four per thousand needlesticks. I'd like to address one question because I must say I disagree with part of Dr. Bauer's presentation. I think the message may have gotten mixed. This business of ascribing risk to areas of the country I don't find in the NCCLS standards and that is not universal precautions. Universal precautions, which are the old blood/body fluid precautions, is a system of barrier precautions that is based on the assumption that one treats all patients in the same basic manner and not the extremes. I can't really put pen to paper and say "How would you do it differently?" In the laboratory how would you treat that blood specimen differently if you knew it was from an AIDS patient and if you knew it wasn't? I've seen this in action in a clinical laboratory in Atlanta, Georgia, which handles about 5,000 serums per day for all sorts of chemical, biochemical and microbiological assays. If a specimen has the word "hepatitis or "dialysis unit" written on it then that specimen goes over in the corner and they're processed under a biological safety cabinet by a person in a full "bunny suit", gloves, gown, etcetera. That's not where the disease is being transmitted. Where it's being transmitted is where the 97% of the rest of the specimens are being handled. And in that particular laboratory no gowns were worn, no lab coats, and people were handling these specimens in little tubes with bare hands and obviously getting contamination on their hands. Now there you have the two extremes. My point is: universal precautions is basically targeted to the middle and that's targeted to prevent hepatitis B. And, if that's correct, and if these procedures are sufficient to prevent hepatitis B then they become very conservative precautions for HIV because of one single factor and that's the difference in titer.

AUDIENCE QUESTION: We have in our city in our teaching hospital a long history of a very low incidence and prevalence of hepatitis B, and I'm sure North Dakota also does. In the eighteen years of being hospital epidemiologist and looking at the issue as carefully as I could the only instance of transmission of hepatitis B to health care workers that we had was a circumstance where a patient with a graft loosened the graft in the surgical ICU and literally showered people with his blood. We knew ahead of time that he was hepatitis B positive. We didn't have any vaccine at that time. Four of those workers came down with hepatitis B and they all did well, fortunately, and none of them became carriers. We've had no other circumstance. We have a very limited number of known AIDS patients and I believe very few HIV-positive patients. Based on Dr. Bauer's data it appears to me that it would take us something like 100 years to come up with one health care worker converting from a needlestick injury. Based on some data from the NIH in terms of looking at other kinds of blood exposure to mucous membrane, hands, and so forth and extrapolating from the known data on HIV, it would take us somewhere between 1,000 and 10,000 years to have one employee convert to HIV positivity. My question is "Does it make any economic sense or does it make any education/inservice sense to focus on the use of gloves for starting I.V.'s and doing phlebotomy for people that are skillful, nurses and med techs that are skillful, and have not demonstrated any risk of hepatitis B in a low hepatitis B/HIV area?" It would probably make more sense to concentrate on issues of needlestick injuries and handwashing which is also important for all those other organisms in the hospital which might get ignored in our concern about "FRAIDS" - the really important epidemic - not AIDS. "FRAIDS" is the hysteria about AIDS. That's the concern I have. We can't get gloves anyway; we can't get the consistency, the good quality gloves. So that's why I am concerned that the glove focus of universal precautions in an institution like mine is a colossal waste of time, money, effort, and focus. Would you like to comment on that?

ANSWER (Dr. Favero): Let me comment briefly, and I'll comment in two ways. One is, as I interpret the CDC recommendations and as I interpret the NCCLS recommendations based on what Stan's said pertaining to gloves, I agree with you. I think there's been a tremendous overreaction to glove use. I agree with everything you say. From the CDC standpoint, since these are recommendations they're very broad and in the body of those recommendations you will always see the phrases that this should be tailored to your own individual setting. And there are infection control strategies that differ. They differ in laboratories, relative to using gloves, they differ in a dialysis unit and they're going to differ on a ward. The persons that are knowledgeable about the setting ought to be the ones that make the decision. I think that as time goes cn and as OSHA gets straightened out on this we're going to come to that. But I agree with you. I think we're usirg way too many gloves in this country. I guess I disagree with my colleague Stan in that I don't think everybody in the lab needs to wear gloves. Do you need to wear gloves to do water microbiology, or if you're streaking stools for Salmonella? I don't think so. I think if you're

handling blood and there's a chance [of contamination], okay, wear a
glove. And wear a glove that's comfortable. And that's not going to
produce a rash. I think we should have been doing that all along.

ANSWER (Dr. Bauer): I said that everybody in the lab should wear gloves.
I mean that everybody in my lab should wear gloves. I think that if you're
in a low-prevalence area I would certainly agree with Marty and you may not
have to wear gloves. Certainly not for low-risk situations. I thoroughly
agree that we all need to have some common sense and I hope that OSHA
recognizes that it's going to have to modify its enforcement criteria to
fit the situation locally.

AUDIENCE QUESTION: I have a question for Robyn Gershon. You mentioned
[that you will be doing] a study of 1600 workers in the Baltimore area.
Will that include regular hospital workers rather than the [HIV production
lab] workers in your other study who have a high degree [of exposure]?

ANSWER (R. Gershon): Yes, they will be hospital workers - doctors, nurses,
technologists, clinical lab techs.

AUDIENCE QUESTION: In my observations, and I'm sure those of many of us in
this room, many health care workers put gloves on and don't change them.
In your study it might be nice to ask "Do you change your gloves, and
when?" because that would be useful data. For phlebotomists, I've recently
seen an update of recommendations that says you don't have to change your
gloves between patients so long as they're not blood-tinged. That probably
is what's happening anyway. Could you comment on that? I see people
working in the laboratory picking up the phone and handling the computer.
I know some people have overcome that by use of dedicated phones and
dedicated computer terminals etcetera, but is that really necessary?

ANSWER (R. Gershon): Certainly in the biosafety level 3 labs everything is
dedicated. Once something goes in it never comes out again unless it takes
a shower out. But I think what you're saying is quite true. I think
people get used to wearing the gloves and they forget they're even on and
they go about their regular activities with the same pair of gloves on.
It's a problem. ["Do you change your gloves?"] is a good question to ask
and I'll make sure we [put it in our study]. We have 14 pages of
questionnaire and I was really hesitant to add any more but now that we
know what the key questions are we can modify them.

AUDIENCE COMMENT: In the entire time that I've been in microbiology we've
treated infectious diseases as a group of diseases, we've adopted policies,
we've thought about them carefully, we've gone through governmental
agencies and private organizations. And we have never before singled out
one disease as being unique for regulatory policy. I'm appalled that we
are now doing this.

PANEL RESPONSE (Dr. Kleger): To comment on your comment that you find
focusing in on HIV "appalling" that was part of my answer about the North
Dakota story. I would like to consider "infectious diseases" rather than
focusing in on HIV. I think that narrows our focus too much. We don't
have much HIV in Fargo but we do have hepatitis in Fargo. We don't have
much HIV or hepatitis in Milwaukee but there are other body-borne diseases
that we have to be concerned about. I don't think we can be cavalier
because we don't have a [high] prevalence of HIV. Dr. Pike has been
studying laboratory-acquired infections for many, many years but only
recently have we started talking about laboratory infections with HIV.
There have been other things all along--ubiquitous infections that exist in
the population and are asymptomatic. The fact is, we've become conscious
of them in the era of HIV because we have government support. But I don't

think we necessarily have to focus in on that one to the exclusion of all others. The same thing is with gloves. Gloves are only part of the story. Handwashing is not ignored in universal precautions or in the NCCLS document.

AUDIENCE QUESTION: I'm a practicing hospital epidemiologist. One cannot lose sight of the fact that there are clear differences between people in the ward taking care of patients and people in the laboratory. You have to keep that in mind when you adopt whatever you do for universal precautions regardless of whether you're in North Dakota or New York City. There are other reasons for universal precautions. Feces may not have HIV or hepatitis B in it but it sure has E. coli. There are other reasons for wearing gloves. So please don't put these aside. Regarding the glove issue, recent data published by Wenzel in the Annals of Internal Medicine suggest that not only are gloves a convenient vehicle for carrying organisms and are resistant to being washed but the organisms also penetrate the gloves very well. So it is necessary to take the gloves off and wash your hands. I think we need more scientific data like this. I do have a question about the assessment of risk in [Robyn Gershon's study]. Are your denominators accurate? When you say the risk of [HIV] infection in various settings, for example, four per thousand health care workers, doesn't that really assume four per thousand needlesticks? One has to consider what the potential for needlestick injury is to arrive at the probability.

ANSWER (R. Gershon): With respect to what the denominator is you probably know that the CDC needlestick survey is essentially needlesticks--that's the definition. But Gerberding and many of the other studies, I think there are eleven now, pooled together make that denominator number take into account mucosal or direct contact exposures. Those are mixed. It's very hard to tease that data out.

AUDIENCE QUESTION: Don't you think that when you're telling someone what the risk is you should be somewhat more legitimate and tell them that their risk may be a lot less than four per thousand?

ANSWER (R. Gershon): It's four per thousand needlestick injuries/exposures and that's exactly what we told them.

AUDIENCE QUESTION: But what is the risk of getting a needlestick? Don't you have to consider that when you're telling someone what their ultimate probability of infection is?

ANSWER (R. Gershon): [The risk of needlestick] varies where you are - it depends on your institution. I think [health care workers] need to make that determination themselves because they know what their own risk is. That's a very hard figure to come up with. It depends, as Stan pointed out, on where you are. Would [Dr. Bauer] like to address that?

RESPONSE (Dr. Bauer): In that calculation I showed earlier I did take that into account and that figure of 25 [needlesticks] per 100 beds per year I think may be the national average. I also think it would be a good performance in a hospital. My own personal feeling is that not more than a third of the real needlesticks occurring in the institution get reported so the numerator data in this particular case is underestimated also.

AUDIENCE QUESTION: My question is for Dr. Bauer. If you test the serum from a laboratory exposure and it comes back positive what do you do with that [result] if you haven't asked the patient if it's okay to test? We require permission for the test to be done. So far we've had two patients come up positive.

ANSWER (Dr. Bauer): The fact is, if a laboratory accident occurs, many laboratories will simply test the serum. We haven't faced that yet. If a housestaff officer is exposed he might simply come down to the lab and have his lab tech friend dig out that patient's serum from the refrigerator and test it. I don't know if that's happening or not. Do you require permission for hepatitis B and RPR's? Why for HIV and not the others?

QUESTIONER RESPONSE: No, only for HIV. Because our risk manager says if we didn't [require permission] we could get into trouble.

RESPONSE (Dr. Bauer): Trouble, but not scientific trouble. You might get into emotional trouble but we're supposed to be scientists and I see no scientific reason not to test RPR, or HBV, or HIV.

AUDIENCE QUESTION: I'm working "in the trenches" doing infection control and trying to deal with the recommendations between OSHA and CDC. In the lab we require all the personnel to wear gloves but we're having a problem with dermatitis from the gloves. One of the employees reacts with everything - underliners, plastic gloves, everything. We were wondering if the dermatitis came under Worker's Compensation because we do require the gloves to be worn. What do we do in that type of situation?

ANSWER (Dr. Bauer): Get her flock-lined plastic gloves that you get in the supermarket. They don't cause dermatitis.

QUESTIONER RESPONSE: We tried those and she reacted to them.

ANSWER (Dr. Kleger): I don't know if this was discussed at NCCLS. It used to be derigueur for society women to wear white cloth gloves. I wonder if they could be worn under latex gloves?

QUESTIONER RESPONSE: She was using [the cloth liners] and they didn't work. We've also had nurses complain about breaking out with rashes. You almost have to wash your hands when you take the gloves off because of the talcum powder.

AUDIENCE QUESTION: We're instituting blood clean-up kits at our hospital. Must urine spills from bags or bedpans be cleaned up the same way as a blood spill according to the latest update from CDC?

ANSWER (Dr. Bauer): CDC says urine is not covered by universal precautions so I think the answer is no. However, if there's blood in the urine then it falls under universal precautions.

AUDIENCE QUESTION: With regards to the Burroughs Wellcome study on HIV needlesticks: as I understand it, that study is a double-blind controlled study with AZT vs. placebo. You will need thousands of health care workers to be enrolled in that trial to learn anything. I think it's a ridiculous study. Although we do offer AZT to our workers who get needlesticks.

RESPONSE (R. Gershon): I agree with you about the trials. I know the lab workers out in the trenches are saying "If we have a stick we do not want to receive the placebo - we want the real thing". [Some] have stockpiles of AZT in their refrigerators and they go right ahead and self-medicate themselves.

AUDIENCE QUESTION: Before I moved to New York City I lived in Iowa. There are blood banks in Iowa which, since the initial screening of Hepatitis B surface antigen was done in the early 1970's, have never had a single unit positive for hepatitis B surface antigen. Despite the fact that the prevalence of surface antigen in the blood bank population of Iowa is extraordinarily small, we had [in our hospital] a seroprevalence of anti-HBs equivalent to most large, metropolitan inner-city hospitals. So I don't think we can be cavalier and say "We don't have hepatitis B here" because you don't see it in the community. Because anybody with hepatitis B is going to end up in the hospital sooner or later. I find it incredible that we can spend a whole morning talking about HIV and universal precautions and not anyone has mentioned hepatitis B vaccine. It's been incredibly under-utilized because people don't push it and administrations in hospitals don't push it because it costs money. The only way you're going to prevent hepatitis B in the healthcare delivery system is to utilize the vaccine because people don't recognize needlesticks. Of healthcare workers who get hepatitis B, [studies have shown] most of them don't remember the needlestick or the patient it occurs from. I'd like to know why people aren't talking about the [hepatitis B] vaccine.

ANSWER (Dr. Kleger): It's one of the topics for the afternoon.

ANSWER (Dr. Willett): I'm glad you brought up that point. In continuing education courses and to the [dental] students we emphasize that the real problem is not HIV, it's HBV. You can't emphasize that enough. We have a problem with the [hepatitisB] vaccine. The students are anxious to get the vaccine but faculty members [deny] they'll ever get hepatitis. I frequently hear "I've been a dentist for 30 years and I'm never going to get hepatitis."

AUDIENCE COMMENT: I have a couple of observations. Unlike most of you here I'm a biosafety officer and my observations are dealing with the fact that what's necessary to protect people in the workplace especially against biological material, is not a new science. It's been around for a long time. My first observation is that I'm surprised given all that we know about what's necessary to protect people, like no eating, smoking, or drinking in the laboratory, which have been discussed for 20 years [these practices] are still being done.

The second comment I'd like to make is that we have to be very careful in how we try to interpret universal precautions at our own institutions until such time as all of the interested parties, CDC, OSHA, NIH, NCCLS, etcetera, get together and come up with some language that allows us or requires us to develop programs that are applicable to our institution and give us the legal latitude before they codify things and preclude options. I would hate to be the person who directs that Iowa facility, who says "You don't have to wear gloves because we don't have a high prevalence rate in our community" and then have one of the workers get a needlestick and then seroconvert. I cannot defend that decision in any court in the United States. So I think that we have to be very concerned and demand from agencies that are going to promulgate regulations that are going to affect how we practice medicine, how we practice dentistry, and how we practice science to really think about what's happening. I've thought about the issue to glove or not to glove for a long time. We have populations of individuals who have been working with radiosotopes for long periods of time. Think back... does anyone have knowledge about people whose hands have been contaminated with radioisotopes if they've used gloves? The answer is no. There's no literature that deals with this. Occasionally if you have a cut or a break in the glove you'll have this problem. There are occupations and there are activities where we don't need gloves. We don't need gloves if we're not dealing with materials that pose risk. If materials pose risk then we at least owe it to our workers to try to protect them in those particular settings. It's not my decision whether to use gloves or not. It's their decision whether their job, in their minds, can be done more safely by them if they're wearing gloves.

AUDIENCE COMMENT: We were talking [earlier] about what we do with needlesticks. I'm an infection control practitioner at a V.A. hospital. [Since] all blood is potentially infectious we follow up the source [for all needlesticks] for HBV and HIV. And we also get a baseline on our employee. Following Dr. Bauer's lecture this morning [there was] some confusion between written consent, informed consent and oral consent. Public Law 100-322 was signed by President Reagan in May. It states that for HIV testing you will have a written documented consent. At this point it has not been published as it must appear in the Federal Register first. They anticipate [it will appear in the Federal Register] in December. Therefore, there is no choice. At this point if you do not have a consent form available then you are supposed to have your source patient sign the progress notes that he has received pretest counselling and has agreed to HIV testing. [This should be done] whether you choose to test him because of his symptoms or following a needlestick. You must have written, informed consent.

MODERATOR RESPONSE: That regulation hasn't gone into effect and I doubt that it ever will. I think that there will be so many people lobbying against it. There are several states that have already passed laws that say you can test and you don't have to get informed consent or written consent if someone sticks themselves with a needle. I personally agree with these laws. I think we should not allow a patient source to refuse [testing] especially if they're at risk [for HIV]. Even though CDC says you follow-up your employee regardless of test results that employee still wants to know what that patient's status is. It will make them feel a lot better if they know the test is negative—and if it's positive they'll know for sure. Even though we know the test isn't 100% accurate.

AUDIENCE QUESTION: This is one for Dr. Willett. How do you suggest that one diplomatically approach a dentist or dental hygienist who refuses to wear gloves during treatment?

ANSWER (Dr. Willet): I'm not sure you should be diplomatic! You can ask him point blank, "There's been so much talk about wearing gloves, how come you're not wearing gloves?" And if he doesn't wear gloves you have a choice: don't go to that dentist! As I pointed out before, sometimes patients will ask "Why are you wearing gloves?" I don't think you hear that as much anymore. At one time people wondered why a dentist needed gloves. But now they know.

REFERENCES

1. Marcus R, CDC Cooperative Needlestick Surveillance Group, Surveillance of health care workers exposed to blood from patients infected with the human immunodeficiency virus, N Engl J Med, 319: 1118-1123 (1988).
2. Centers for Disease Control, Acquired immune deficiency syndrome (AIDS): precautions for clinical and laboratory staffs, MMWR, 31: 577-580 (1982).
3. Centers for Disease Control, Acquired immunodeficiency syndrome (AIDS): precautions for healthcare workers and allied professionals, MMWR, 32: 450-451 (1983).
4. Conte JE, Hadley WK, Sande M and the University of California, San Francisco Task Force on AIDS, Infection control guidelines for patients with the acquired immunodeficiency syndrome (AIDS), N Engl J Med, 309: 740-744 (1983).
5. Centers for disease control, Recommendations for preventing transmission of infection with human T-lymphotropic virus type III/lymphadenopathy-associated virus in the workplace, MMWR, 34:681-695 (1985).
6. Centers for Disease Control, Recommendations for preventing transmission

of infection with human T-lymphotropic virus type III/lymphadenopathy-associated virus during invasive procedures, MMWR, 35: 221–223 (1986).

7. Centers for Disease Control, Recommended infection control practices for dentistry, MMWR, 35: 237–242 (1986).

8. Gerberding JL, the University of California, San Francisco Task Force on AIDS, Recommended infection control policies for patients with human immunodeficiency virus infection: an update, N Engl J Med, 315: 1562–1564 (1986).

9. Centers for Disease Control, Recommendations for prevention of HIV transmission in healthcare settings, MMWR, 36: suppl 2S (1987).

10. American Hospital Association, Statement of the advisory committee on infections within hospitals on protection of health care workers, Chicago: AHA (1987).

11. Lynch P, Jackson MM, Cummings MJ, Stamm WE, Rethinking the role of isolation practices in the prevention of nosocomial infections, Annals of Internal Medicine, 107: 243–246 (1987).

12. Garner JS, Simmons BP, Guideline for isolation precautions in hospitals, Infection Control, 4 (suppl): 245–327 (1983).

13. U.S. Department of Labor and U.S. Department of Health and Human Services, Joint advisory notice: protection against occupational exposure to hepatitis B virus (HBV) and human immunodeficiency virus (HIV), Federal Register, 52: 41818–41824 (1987).

14. Occupational Safety and Health Administration, Occupational exposure to hepatitis B virus and human immunodeficiency virus, Advance notice of proposed rulemaking, Federal Register, 52: 45438–45441 (1987).

15. Occupational Safety and Health Administration, OSHA instruction CPL.2-2.44: Enforcement procedures for occupational exposure to hepatitis B virus (HBV), human immuncdeficiency virus (HIV), and other blood-borne infectious agents in health care facilities. January 19, 1988.

16. Centers for Disease Control, Update: Universal precautions for prevention of transmission of human immunodeficiency virus, hepatitis B virus, and other bloodborne pathogens in healthcare settings, MMWR, 37: 376–388 (1988).

17. Occupational Safety and Health Administration, OSHA instruction CPL 2-2.44A: Enforcement procedures for occupational exposure to hepatitis B virus (HBV) and human immunodeficiency virus (HIV). August 15, 1988.

INFECTIOUS WASTE TREATMENT AND DISPOSAL ALTERNATIVES

Lawrence G. Doucet

Doucet and Mainka, P.C.
Consulting Engineers
2123 Crompond Road
Peekskill, NY 10566

INTRODUCTION

Infectious waste management and disposal issues are of prominent national concern. Widely reported illegal disposal incidences and beach washups over the last few years have stirred public fears and anger. Politicians and legislators have responded by enacting stringent legislation for the management, manifesting and disposal of infectious waste.

As a result of recent legislation and guidelines on a state and national level, as well as other concerns, the quantities of waste to be managed and disposed as potentially infectious at many hospitals and other institutions have increased enormously. At some hospitals, "infectious" waste quantities have increased from a level of about 3 percent of total solid waste to nearly 90 percent of total solid waste. Such rapid and voluminous increases have created severe difficulties for many hospitals to locate or select reliable, safe and cost-effective alternatives for treating and disposing of their infectious waste.

On-site infectious waste treatment technologies, such as steam sterilization, shredding with chlorination, incineration, and off-site disposal services have comparative advantages and disadvantages which substantially affect their viabilities on a case-by-case basis. Technological, environmental, regulatory, economic and socio-political factors must all be carefully considered prior to selecting and implementing one of these alternatives.

INFECTIOUS WASTE GENERATION

The first and most important step in evaluating and selecting an infectious waste treatment and disposal program is to identify or define the types, sources and quantities of waste which require management and

disposal as potentially infectious. Such definitions must not only
consider present generation rates but also potential future increases due
to policy or regulatory changes. Inaccurate estimates or projections could
result in the procurement of a waste treatment system of either inadequate
capabilities or which is excessively complex and costly.

There are five primary factors which influence or determine the
quantities of waste which require treatment and disposal as potentially
infectious. These are:

1. Regulatory Definitions and Guidelines

    Federal, state and local designations and definitions for "infectious"
    waste vary widely and are sometimes vague and ambiguous.

2. Interpretations of the Regulations and Guidelines

    Site-specific and individual interpretations of regulatory
    definitions, and the intentions of such definitions, can substantially
    affect infectious waste generation rates. Regulatory agencies and
    individual institutions may have widely divergent opinions as to which
    waste stream components and sources need to be regulated as
    infectious.

3. Waste Management Policies and Protocols

    Individual hospitals and other institutions establish protocols and
    procedures for segregating and managing infectious waste in compliance
    with regulatory and accreditation requirements. The conservatism of
    such policies also varies widely. For example, many hospitals have set
    a policy whereby all patient-contact waste is considered potentially
    infectious in line with their own interpretations of CDC "universal
    precaution" guidelines.

4. Waste Management Program Effectiveness

    The ability to implement and effectively administrate an infectious
    waste segregation program can substantially impact the quantities of
    waste requiring treatment and disposal as "infectious" waste.
    The best protocols and written procedures are no better than the
    personnel assigned to implement them. Sloppy and unsupervised waste
    handling and packaging procedures could easily result in large
    quantities of (non-infectious) trash being intermixed with infectious
    waste items. Likewise, infectious waste items could also be
    inadvertantly intermixed with general trash and cause other problems.

5. Off-Site Haulage and Disposal Restrictions

    This factor is the most significant of all. Regardless of regulatory
    requirements or other in-house programs, local restrictions or
    prohibitions by general waste haulers, sanitary landfill operators or
    municipal waste incineration facilities can effectively result in all
    of the waste from a hospital being considered "infectious." There is
    little recourse should the municipal waste transporters and disposal
    firms within a municipality or region refuse to handle or dispose of
    any hospital waste by unilaterally claiming that all of it is
    "infectious." This has happened in several municipalities.

The quantities or generation rates of infectious waste resulting from the above factors are site-specific. The variations can range from as little as 3 percent of total solid waste to as much as 100 percent of total solid waste. Table 1 shows typical, approximate ranges for these factors.

TABLE 1

INFECTIOUS WASTE GENERATION COMPARISONS

TYPICAL PARAMETERS	APPROXIMATE INFECTIOUS WASTE GENERATION RANGES (PERCENTAGES OF TOTAL WASTE)
● Centers for Disease \a Control (CDC) Designations	3 - 5%
● U.S.EPA Guidelines \b	7 - 15%
● Designated Departments \c	20 - 35%
● All Patient-Contact Waste \d	60 - 90%
● Hauler/Disposal Facility Restrictions	0 - 100%

\a   Reference 3

\b   Reference 7

\c   Proposed U.S.EPA RCRA, Sub-Title C, Hazardous Waste Regulations, 1978; Departments include Autopsy, Emergency, ICU's, Isolation Rooms, Clinical Labs, Obstetrics (including patient rooms), Pathology, Pediatrics & Surgery (including patient rooms)

\d   Based upon site specific - interpretations of "Universal Precautions" per CDC "Recommendations for Prevention of HIV Transmission..." Morbidity & Mortality Weekly Report, Vol. 36, August 21, 1987

INFECTIOUS WASTE TREATMENT TECHNOLOGIES

The only proven technologies for treating and disposing of the large and increasing quantities of infectious waste being generated at many hospitals and other institutions are steam sterilization, or autoclaving, shredding with chlorination and incineration. Other technologies, such as dry heat sterilization, gas/vapor sterilization and radiation are either too limited in capacity or are unproven for processing large waste volumes. Innovative or emerging technologies, such as glass slagging systems, high-temperature plasma systems and systems combining shredding and radiation are in the development stage and are years away (if ever) from being proven or made commercially available. Alternative infectious waste treatment technologies are shown on Table 2.

TABLE 2

ALTERNATIVE INFECTIOUS WASTE TREATMENT TECHNOLOGIES

● STEAM STERILIZATION

    -- Gravity Systems
    -- Pre-Vacuum Systems
    -- Retort Systems
    -- Combination Trash Compactor/Autoclave Units

● SHREDDING/CHLORINATION

    -- Small Scale Sharps/Lab Waste Processing Systems
    -- Large Scale Total Infectious Waste Processing Systems

● INCINERATION

    -- Multiple Chamber Systems
    -- Controlled Air Systems
    -- Rotary Kiln Systems
    -- Innovative Systems

● OTHER (SMALL SCALE) SYSTEMS

    -- Dry Heat Sterilization
    -- Gas/Vapor Sterilization
    -- Radiation

● EMERGING TECHNOLOGIES

    -- Glass Slagging
    -- High-Temperature Plasmas
    -- Shredding/Radiation
    -- Etc.

The principal, proven technologies are as follows:

1. Steam Sterilization

    Autoclaving basically involves a system whereby steam is brought into
    contact with waste materials in a controlled manner and for sufficient
    duration to kill pathogenic micro-organisms which may be contaminating
    the waste. The different types of autoclave systems and designs
    relate to steam contact efficiencies and waste volumes which can be
    processed, or sterilized, within the shortest possible time periods.
    Sterilization performance, or efficiency, is a direct function of
    steam penetration into the packages of waste being treated by the
    system. Factors such as waste type and density, packaging materials
    and waste loading procedures directly affect steam penetration and the
    exposure times necessary for effective sterilization. Inadequate
    steam penetration is usually the limiting factor in achieving
    sterilization within a reasonable time period.

    In systems whereby steam pressure alone is used to evaluate air from
    the autoclave chamber, termed gravity systems, only about 15 minutes
    of direct steam contact is typically required with steam temperatures
    of about 250 F, which is equivalent to about 250 psi of steam
    pressure. However, actual cycle times for gravity systems are usually
    about 60 to 90 minutes (per load) in order to allow for full steam
    penetration into the most densely packed waste loads. Other designs
    using vacuum pumps to evacuate air from the chamber, termed pre-vacuum
    systems, have more rapid and efficient steam penetration. Therefore,
    cycle times for pre-vacuum systems range from only about 30 to 60
    minutes (per load).

    Retort type autoclave systems, basically comprise large volume chambers
    designed for high steam pressures, and hence, minimal cycle times. At
    least one commercial disposal firm on the west coast uses retort
    autoclaves for treating infectious waste.

    A unique, autoclave system variation features a combination, integral
    pre-vacuum sterilizer and general waste compactor. After the
    autoclave cycle is completed, sterilized infectious waste is
    automatically ejected into the trash compactor section. All treated
    waste and trash is then compacted into a close-coupled, roll-off type
    container for off-site disposal.

    Two types of packaging are viable for autoclaving infectious waste as
    follows:

    ●   Heat-resistant autoclavable bags, typically made of polypropylene
        plastic, which are sturdy and will not melt at steam
        temperatures. This type of bag needs to be opened prior to
        autoclaving to allow steam penetration into the waste.

    ●   Heat-sensitive, low-density polyethylene bags which will melt at
        steam temperatures. Such melting facilitates steam penetration
        and air evacuation. The use of these bags requires secondary
        containment to prevent spillage of waste from the melted bags.

    In order to assure that the autoclave systems are loaded, operated and
    maintained properly, temperature probes and frequent biological
    challenging are required. Biological challenging involves the
    insertion of heat resistant spore samples (such as bacillus

stearothermophillus) into worst-case waste loads in order to monitor and verify sterilization efficiencies. Some states have imposed stringent requirements for such challenging and monitoring.

The principal advantages of steam sterilizations systems include low capital and operating costs, relatively small space requirements and simplicity of operations.

The principal disadvantages of steam sterilization systems are that they have relatively limited capacity, may require special waste packaging and handling and need special provisions to prevent odor and drainage problems. Autoclaving is not recommended or suitable for all wastes, including pathological waste, such as carcasses and body parts, high liquid content waste, such as bulk fluids and blood, and waste contaminated with volatile chemicals, such as chemotherapy waste.

A potentially serious problem of using autoclaves to treat infectious waste is that of disposing of the treated waste. After autoclaving, waste appearances are basically unchanged, and color-coded bags and international biohazard symbols remain intact and visible. Needles, syringes, IV tubing, red colored and blood stained waste items and the like may be totally sterile, but are still recognizable and possibly not acceptable for disposal with general waste. Compacting autoclaved infectious waste tends to break open waste bags and other containers and expose and spill their contents. Consequently, waste haulers and landfill operators may not accept autoclaved waste even if they are proven to be sterilized.

2. Shredding with Chlorination

Within the last few years, a technology featuring a combination of shredding and chemical sterilization has been widely promoted by a midwestern firm. Two models are available. One is designed for relatively small and limited quantities of laboratory waste and sharps. The other is a relatively large capacity system designed for treating almost all infectious waste generated in a hospital.

With the large capacity system, waste is manually loaded onto an inclined, conveyor belt which feeds a high-torque, low-speed shredder. Waste is discharged from the bottom of this shredder into a high-speed hammermill which granulates the waste. During both shredding stages, waste is continuously sprayed and saturated with sodium hypochlorite solution. An inclined, perforated conveyor at the discharge of the hammermill separates the granulated waste, or debris, from the excess liquids, or slurry. The slurry is collected in a basin and piped to a sewer drain, and the solids are discharged into a cart where they are retained for off-site disposal. Reportedly, sodium hypochlorite contact time in the system and cart is sufficient for sterilization.

The sodium hypochlorite can be generated on site in an electrolysis process from water and salt pellets, or it can be purchased in bulk quantities. The standard, or base, generator furnished with the system requires about 24 hours to generate sufficient sodium hypochlorite solution for about 90 minutes of operation.

To prevent airborne contamination from the process, a blower draws air from the discharge hoods of the feed and debris conveyors and maintains a negative pressure on the entire system. The air passes through a series of prefilters and a (chlorine resistant) HEPA filter before being discharged to atmosphere.

The principal advantages of shredding/chlorination systems are that they are relatively simple, provide a substantial volume reduction, alter the waste appearance and form such that all items are unrecognizable and are suitable for most types and forms of infectious waste, except pathological remains. Hourly processing capacity is about 800 to 1,000 pounds, but it reportedly can be as high as 2,000 pounds. Daily throughput is a function of system sodium hypochlorite generation capacity or purchase. Waste volume reduction is estimated to be about 5 to 1, but it reportedly can be as high as about 8 to 1.

The principal disadvantages of shredding/chlorination systems are that they have relatively high costs, relatively limited throughput capacities and potential problems with slurry contaminants, workplace chlorine concentrations and noise levels. A standard, large capacity system can cost as much as a small to medium capacity incineration system. The slurry discharged to sewer may have concentrations of metals, organics and other contaminants such that a discharge permit may be required. In addition, special precautions may be needed to assure compliance with occupational workplace standards and requirements.

Another important consideration is that shredding/chlorination systems are currently only offered by a single manufacturer, and only a single, large capacity, operational system is currently in existence. This larger system is installed at a midwestern hospital which incinerates most of its waste on-site. However, it has been reported that two large capacity systems have been purchased by the Ontario Ministry of the Environment for demonstration purposes. Also, it should be noted, that there are several, reportedly successful small capacity systems in operation.

3.   Incineration

Incineration is basically a process using controlled, high temperature combustion to destroy organics in waste materials. Modern incineration systems are well engineered, proven, high technology processes designed to maximize combustion efficiencies and completeness with minimum emissions.

There are four basic hospital/institutional waste incineration technologies suitable for disposing of infectious waste. These are as follows:

1.   Multiple-Chamber Incinerators

This technology was developed in the mid-1950's and it was virtually the exclusive type of hospital/institutional waste incineration system installed through the mid-1960's. This type of system is also termed Incinerator Institute of America, or IIA, technology. Multiple-chamber incineration processes are designed for very high excess-air levels and have settling chambers in order to control combustion and help limit emissions. However, virtually all of these systems need air pollution control devices in order to comply with emission regulations. In addition, they cannot meet the current performance and operating requirements in many states without substantial upgrading and the addition of state-of-art combustion control equipment.

Very few multiple-chamber incinerators are being built today, but almost all of the existing incinerators that are more then 25 years old are of this type. The smaller capacity systems feature solid hearths which were strictly designed for burning pathological waste. Many hospitals have attempted to use these small capacity, solid hearth pathological incinerator systems for burning infectious waste. However, severe operating and emission problems usually result from this type of misoperation. Some of the other larger capacity, multiple-chamber units were built with grates in the solid waste (primary) burning chamber. When infectious waste is burned in these systems, uncombusted waste materials fall through the grates into the ash pit. Operators are exposed to potential hazards when cleaning infectious items and sharps from the ash pits under the grates.

2. Rotary Kiln Incinerators

A rotary kiln incinerator basically features a cylindrical, refractory-lined, combustion chamber which rotates on a slightly inclined, horizontal axis. Waste is loaded at one end of the kiln, and the rotation moves the waste slowly towards the opposite end where it is discharged as ash. The kiln rotation helps promote good burnout and a superior ash quality. Rotary kiln systems require secondary combustion chambers and air pollution control equipment in order to comply with emission regulations.

Rotary kiln incinerators are widely used in industrial applications for burning hazardous waste. This is largely because the technology is very versatile and suitable for most types and forms of waste, including solids, sludges, liquids and even fumes. Within the last few years, these systems have been widely promoted for burning hospital waste. However, today there are only about half-dozen rotary kiln incinerators installed in hospitals and similar institutions across the country.

One of the reasons why there are so few rotary kilns installations at hospitals is that they have relatively high capital and operating costs. For comparable capacities, they are roughly twice as costly as other institutional waste incineration technologies. Rotary kilns also have relatively high maintenance and repair requirements because of the abrasive and scraping effects of waste being tumbled and dragged along the refractory lining of the kiln as it rotates.

Another potentially major problem with using rotary kilns for incinerating infectious waste on-site is that, in most applications, the waste is required to be processed as it is being loaded. This is usually accomplished with a special type of loader, termed an "auger feeder," which uses a teethed, screw mechanism, that shreds, crushes and extrudes waste into the kiln. Such processing spills and disperses the contents of infectious waste bags and containers within the feeding mechanism, thus creating potential maintenance and clean-up hazards.

3.   Controlled Air Incinerators

This is also commonly called modular combustion and starved air
incineration. Controlled air incineration is basically a two-
stage combustion process. Solid waste is burned in a starved
air, or reducing, environment in the first stage, or the primary
chamber. Combustion products and volatile gases generated from
the solid waste in the primary chamber are burned under excess
air conditions in the second stage, or secondary chamber.

The first controlled air incinerators were installed in this
country in about 1962. The technology was initially popular
because of its relatively low costs, but its popularity grew
quickly primarily because most systems could readily comply with
air pollution control regulations without needing emission
control equipment. On the order of 7 to 10 thousand controlled
air incinerators have been installed in the last twenty years,
and more than 95 percent of all the hospital/infectious waste
incinerators installed in the past 20 years have been this type
of system. It should be noted, however, that no controlled air
incinerators will be able to comply with the stringent emission
control regulations being legislated in many states without air
pollution control equipment.

4.   Innovative Systems

This type of incineration technology includes a wide range of
"designs," "new" developments, unusual applications and avante
garde systems offered by various "progressive" manufacturers and
promoters. Although many such systems are continually being
"developed" and may appear promising on the surface, the majority
have never been demonstrated in actual operation. Some "designs"
are based upon reincarnations of old failures, and some defy the
laws of physics and thermodynamics. Anyone considering a new
technology or "innovative" system should understand that there is
a wide difference between an idea or conceptual schematic and a
proven application.

An incineration system is an integration of various components of
which the incinerator proper is only a single element. All components must
be properly designed and coordinated to function with the other components
in order for the system to operate successfully. Incineration system
components include waste handling and loading systems, burners and blowers,
ash removal and handling systems, waste heat recovery systems, emission
control systems, breechings and stack systems and controls and
instrumentation.

There have been numerous developments over the last several years
which have improved incineration system operations and efficiencies. For
example, waste loaders have recently been developed for accepting larger
waste capacities. Some newer loader designs can hold as much as an hour's
worth of loading at one time. Burner and blower systems are available with
state-of-art controls and full-integration so as to minimize auxiliary fuel
usage and provide maximum combustion control during the full cycle of
system operations. Modern ash removal systems featuring backhoes and
scoops have been developed which appear to be more reliable and less
maintenance intensive than cart and drag type conveyor systems. Some
manufacturers have developed controls and instrumentation packages with

solid-state programmable controllers, graphic displays and even touch-screens.

The addition of a waste heat recovery boiler to an incineration system is not nearly as cost-effective as it was ten years ago; in fact, nowadays, it is rare for a hospital/institutional waste incinerator to be justified strictly on the basis of heat recovery benefits. On average, about 3 to 4 pounds of steam can be recovered for each pound of infectious waste incinerated. However, at the higher operating temperatures required by many states, about 5 to 6 pounds of steam can be recovered for each pound of waste incinerated. Although such recovery rates are seldom sufficient to provide a rapid return-on-investment for a total system, the addition of a heat recovery system may have other advantages. For example, incineration with heat recovery is usually considered more acceptable, or less objectionable, to the general public than one without heat recovery. Also, a heat recovery system may help to condition flue gases upstream of an air pollution control system. Finally, energy grants may be available for systems with heat recovery.

In many states, new legislation requires that hospital/infectious waste incinerators be equipped with air pollution control systems and equipment meeting "Best Available Controlled Technology," or BACT. Such systems are very sophisticated and energy intensive as needed to achieve <u>extremely</u> stringent particulate and acid gas, or hydrogen chloride (HCl), emission levels. The most proven and widely used emission controls systems applicable to hospital/infectious waste incinerators are high-energy venturi scrubbers with packing sections, sub-coolers, mist eliminators, caustic feeders (pH controllers) and water recirculation systems. Fired reheaters are also available for eliminating visible steam plumes from the stacks of systems with wet scrubbers.

There have been recent attempts to develop relatively small capacity "dry" scrubbing systems which use baghouse filters and alkaline injectors for combination high efficiency particulate removal and (moderately efficient) acid gas removal. However, not only are such "dry" systems nearly twice as costly and space intensive as wet scrubber systems, but also, to date, none have been used successfully on any on-site hospital/infectious waste incineration system in the country. Nonetheless, some state regulatory agencies are essentially requiring that such "dry" systems be installed on all new infectious waste incinerators.

By and large, on-site incineration has emerged as the preferred, most viable infectious waste treatment option for most hospitals and institutions. From a technological standpoint, incineration offers several major advantages as compared to other treatment technologies. More importantly, it may be the only treatment method with a processing capacity suitable for the infectious waste generation rates of most hospitals and other institutions. Incineration not only sterilizes infectious waste but also provides typical weight and volume reductions of 90 to 95 percent. Incineration of total hospital waste minimizes many difficulties and problems associated with the segregation of infectious waste. In addition, it converts obnoxious waste, such as animal carcasses, to innocuous ash, provides the potential for waste heat recovery and, in some cases, can be used for simultaneously disposing of hazardous chemicals and low-level radioactive waste.

On-site incineration attractiveness is also greatly enhanced by various current and pending legislation. About half of the states and several major cities currently mandate that infectious waste be treated on-site, restrict its off-site transport and/or prohibit it from being landfilled. Many additional states are planning similar, restrictive legislative measures within the next few years. Virtually all states either require, recommend or advocate incineration as the preferred method for treating infectious waste. Furthermore, incineration is the only treatment technique recommended in the U.S.EPA <u>Guide</u> for virtually all designated infectious waste types.

Off-site disposal difficulties and limitations probably contribute the greatest incentives for many hospitals and other institutions to consider or select on-site incineration as the preferred infectious waste treatment method. It has become increasingly difficult, if not impossible, to locate reliable, dependable infectious waste disposal service contractors. Many hospitals able to obtain such services are literally required to transport their infectious waste across the country to disposal facilities. Furthermore, such services are typically very costly, if not prohibitive. Off-site disposal contractors are typically charging from about $0.30 to about $0.80 per pound of infectious waste, and some are charging as much as $1.50 per pound. On the other hand the <u>total</u>, annual owning and operating costs for hospital/infectious waste incinerators in states with even the most stringent legislation range from about $0.05 to about $0.20 per pound of waste incinerated. This is inclusive of system amortization costs, utility costs, operating labor, ash disposal, testing and maintenance and repair.

A major disadvantage of on-site incineration, compared to other treatment technologies, are its high capital, operating and maintenance costs. However, more importantly, regulatory restrictions, socio-political opposition and related permitting difficulties have made on-site incineration increasingly prohibitive, if not impossible, in more and more sections of the country. In an effort to protect the environment and public welfare against potentially unacceptable emissions, an increasing number of state and local pollution control agencies are enacting extremely restrictive regulations and criteria for permitting and operating infectious waste incinerators. Unfortunately, many such regulations appear to have no technical basis, and they are often unrealistic and sometimes unattainable.

It is likewise becoming more and more difficult, if not impossible, for permit applicants to prove or otherwise demonstrate that properly designed and operated incineration systems are environmentally benign and pose no significant increased risks. The major reason for such difficulties is public opposition. Issues such as fear, mistrust and the "not-in-my-backyard" (NIMBY) syndrome cannot be effectively countered with scientific data or logic. Consequently, since most regulatory agencies tend to take a passive or neutral position at public hearings, an increasing number of infectious waste incinerator permits are being denied or indefinitely postponed.

Table 3 summarizes the major components of the three principal treatment technologies, and Table 4 summarizes their comparative advantages and disadvantages.

OFF-SITE TREATMENT AND DISPOSAL

There are basically only three options potentially available as an alternative to on-site treatment. These are as follows:

1. Contract Disposal

This involves paying a fee to an independent, commercial firm to
transport and dispose of infectious waste at an off-site facility.
Almost all of the contractors use incineration for disposal. Some
contractors provide waste transport and incinerate at their own
facilities, and others only provide transport and use the incineration
facilities of another contractor. Some disposal contractors have
arrangements to use, or share, on-site incineration facilities at
various hospitals. Contract disposal rates are typically set at a
cost per pound or a cost per box basis. Contractors often furnish
packaging materials and boxes as part of their services. Some offer
refrigerated trucks for longer term, interim storage and transport.

2. Disposal at another Institution's Incinerator

Some hospitals have excess incineration capacity and offer disposal
services to other regional hospitals, clinics and medical facilities.
Such services are on a fee arrangement or shared cost basis which is
typically very competitive with contract disposal rates. Excess
incineration capacity at most hospitals is only incidental to their
existing operations, but at some hospitals it is a planned investment
opportunity.

3. Disposal at a Regional Incineration Facility

A regional incineration facility, as opposed to a contractor owned
facility, is basically developed, owned and operated on behalf of and
under the control of an independent hospital group or
association. An association could develop, administrate and
finance such a facility through either a private developer
or through their own internal organization. The facility
could be either at a neutral site or at the site of a
membering hospital.

The advantages of off-site treatment and disposal include simplicity
and relatively short implementation time. It avoids problems and
uncertainties of siting and permitting an on-site treatment system.
Building space and associated support services are not required. In
addition, off-site treatment and disposal eliminates major capital
investment requirements for on-site treatment facilities.

As discussed, a major difficulty with off-site treatment and disposal
services in many parts of the country is locating reliable, reputable and
affordable contractors and facilities. At present, there is a severe
shortage of off-site incineration capacity on a national level. Most
existing, permitted facilities are operating at peak capacity. Some states
have moratoriums on new, off-site, contract incineration facilities, and,
in the other states contractors are finding it extremely difficult to site
and permit new facilities.

Despite potentially attractive economic incentives, most hospitals are
hesitant or reluctant to incinerate waste from other hospitals. They
appear to have major concerns as to potential liabilities and adverse
neighborhood reactions to such operations. Furthermore, there are few, if
any, operational regional incineration facilities. The implementation of
such facilities has also been stymied by siting and permitting problems.

Also, as discussed, another major disadvantage of off-site transport and disposal, as compared to on-site treatment, are the high annual costs. The costs for off-site, contract disposal are many times greater than those for on-site incineration. The differential is such that many on-site, hospital waste incineration systems realize payback periods of less than 2 years due to off-site disposal cost savings.

The Medical Waste Tracking Act of 1988 (Act) and comparable legislation in many states also impose difficulties and additional, increasing costs for the off-site disposal of infectious waste. Packaging, manifesting and tracking requirements, as well as the severe penalties associated with the violation of the requirements, are significant deterrents to off-site disposal. It has been estimated that the costs for many hospitals to administrate and adhere to the manifesting and tracking requirements under the Act will be greater than those for incinerating their infectious waste on-site. Civil penalties for noncompliance are as much as $25,000 per violation, and criminal penalties are as much as $50,000 and 5 years imprisonment per violation.

A regional incineration facility, as compared to individual, on-site incineration facilities, has the advantage of favorable economics, centralized control and operations and the need to obtain only a single permit. However, as noted, locating a site that can be permitted for incinerating infectious waste with minimal public opposition is extremely difficult and may be comparable to siting a nuclear power plant. In addition, packaging, manifesting and tracking requirements could also have a major impact on the hospitals using the regional facility, even if they own and operate it.

The comparable advantages and disadvantages of the various off-site treatment and disposal alternates are shown on Table 5.

A schematic of the infectious waste treatment and disposal alternatives discussed above is shown on Figure 1.

EVALUATION AND SELECTION

Two steps are recommended for evaluating infectious waste treatment and disposal options and alternatives or for planning a waste management program. These are as follows:

1. Data Collection, Confirmation & Summary

    The initial step is to compile and consolidate all data and information necessary for identifying and evaluating the options and alternatives. Such data typically include waste characterization and quantification, waste handling practices and procedures, site availability and constraints, utility availability, costs for labor, utilities and off-site disposal and the latest regulatory requirements. This also includes a review of in-house policies regarding infectious waste management and disposal, particularly with regard to the likelihood of their being revised in the near and long-term.

    Waste characterization and quantification are the key parameters in formulating a waste management and disposal plan and selecting the best disposal alternative. The efforts required for collecting such data can vary widely, ranging from the use of empirical factors and

TABLE 3
PRINCIPAL INFECTIOUS WASTE TREATMENT SYSTEM COMPONENTS

AUTOCLAVING SYSTEM	SHREDDING/CHLORINATION SYSTEM	INCINERATION SYSTEM
• Waste Transport/Treatment	• Waste Feed Conveyor	• Waste Handling & Loading
• Autoclavable Bags	• Pre-Shredder	• Incinerator
• Autoclave Chamber	• Hammermill	• Burners & Blowers
• Ventilation System	• Debris Conveyor/Separator	• Ash Removal & Handling
• Container Dumper (Optional)	• HEPA Filtration System	• Breeching, Blowers, Dampers & Stack(s)
• Biological/Temperature Indicators	• Sodium Hypochlorite System	• Air Pollution Control
		• Waste Heat Recovery
		• Controls & Instrumentation

TABLE 4

INFECTIOUS WASTE TREATMENT TECHNOLOGY COMPARISONS

PRINCIPAL TREATMENT TECHNOLOGIES	ADVANTAGES	DISADVANTAGES
● AUTOCLAVING	• Low Costs • Low Space Requirements • Ease of Implementation • Simplicity of Operation	• Limited Capacity • Not Suitable for all Wastes • Waste Handling System/Bags • Odor Control • Waste Volume Unchanged • Waste Appearance & Form Unaltered
● SHREDDING/CHLORINATION	• Substantial Volume Reduction • Suitable for Many Wastes • Relative Simplicity • Alters Waste Forms	• Relatively High Costs • Manual Waste Handling • Limited Capacity • Liquid Effluent Contaminants • Room Noise & Chlorine Levels • Single Manufacturer • Limited Experience
● INCINERATION	• Disposes of Most Waste Types & Forms • Suitable for Large Volumes • Largest Weight & Volume Reductions • Sterilization & Detoxification • Heat Recovery	• Relatively High costs • High M&R Requirements • Stack Emissions & Concerns • Permitting Difficulties • Public Opposition

TABLE 5

OFF-SITE INFECTIOUS WASTE TREATMENT & DISPOSAL COMPARISONS

ADVANTAGES	DISADVANTAGES

- **OFF-SITE DISPOSAL**
  - -- Commercial Facility
  - -- Another Institution's Incinerator
  - -- Regional Facility

ADVANTAGES	DISADVANTAGES
• Negligible Capital Investment	• Locating Reliable & Reputable Firms & Facilities
• Minimal (On-Site) Space Requirements	• Potential Liabilities & Concerns
• Simplicity	• High Annual Costs
• Short Implementation Time	• Special Packaging Requirements
• Avoid On-Site Disposal Permitting	• Manifesting & Tracking

- **REGIONAL OR SHARED-SERVICE INCINERATION FACILITY** (vs. Individual On-Site Incinerators)

ADVANTAGES	DISADVANTAGES
• Favorable Economics	• Siting & Permitting Difficulties
• Single Permit	• Special Packaging & Transport Requirements
• Centralized Operations	• Manifesting & Tracking
	• "Hazardous" Designation (some states)

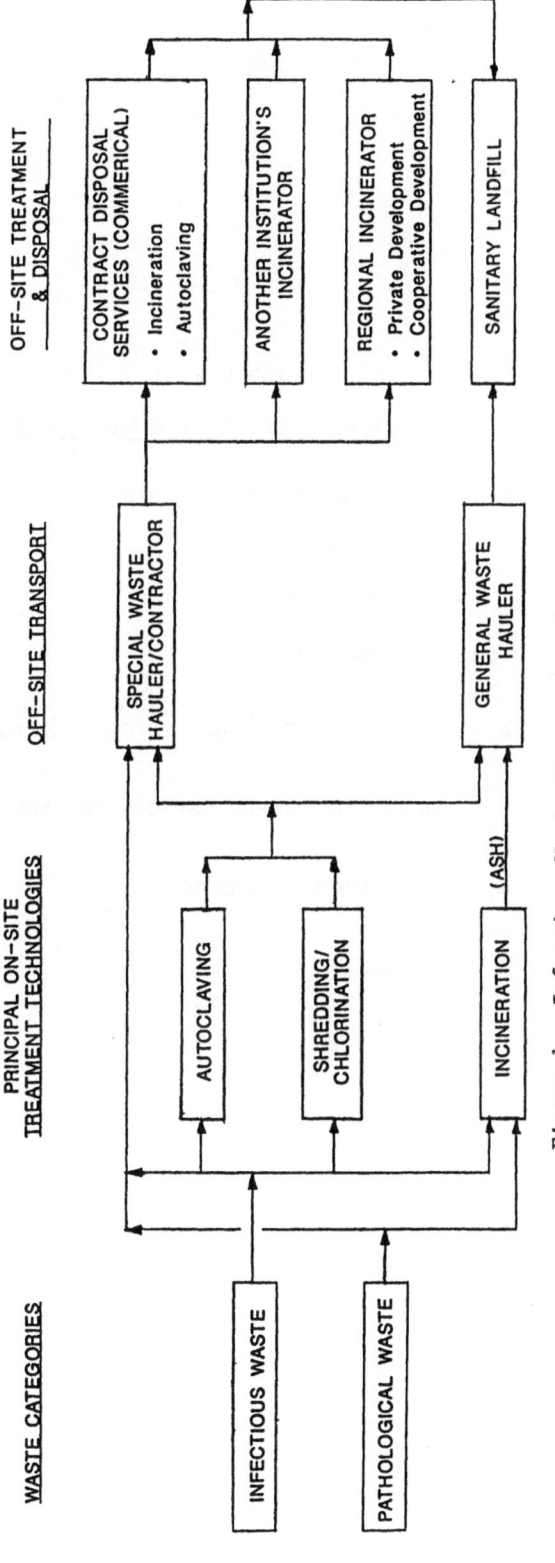

Figure 1. Infectious Waste Treatment and Disposal Alternatives

TABLE 6

TYPICAL DISPOSAL OPTION VARIABLES

- DEGREE OF ON-SITE TREATMENT

  -- None

  -- Selected

  -- Maximum

- ALTERNATE TECHNOLOGIES/COMBINATIONS

- TREATMENT TECHNOLOGY OPTIONS/ADD-ONS

- REDUNDANCY & BACK-UP

- SITING

approximations to the implementation of extensive waste weighing and survey programs. Likewise, such survey programs can range widely in complexity and scope. They require careful planning, organization and coordination to obtain the needed data at minimum cost and effort.

2. Technical and Economic Evaluations

After all relevant data have been compiled and confirmed, this step involves the identification and evaluation of all waste disposal options and alternatives. A typical matrix of options would include varying degrees of on-site treatment, different types of waste treatment technologies and equipment, various potential treatment system add-on features and redundancies. It would not be unusual for a dozen or more viable options to be identified for any particular facility.

Other factors which need to be considered include such parameters as back-up capabilities and contingencies, future facility growth, future waste generation scenarios, potential liabilities and public image.

The important disposal option variables are shown on Table 6.

It is important that the advantages, disadvantages and limitations of the various on-site treatment technologies and off-site disposal options be thoroughly understood. In short, these assessments provide the foundation for ultimately selecting and implementing the best, most cost effective alternative.

BIBLIOGRAPHY AND REFERENCES

1. Centers for Disease Control, U.s. Dept. of Health & Human Services, "Recommendations For Prevention of HIV Transmission in Health-Care Settings," Morbidity and Mortality Weekly Report, Vol. 36, August 21, 1987.

2. Cheremisinoff, P.N., Banerjee, K., Sterilization Systems, Technical Publishing Co., 1985.

3. Centers for Disease Control, U.S. Dept. of Health & Human Services, "Guideline for Handwashing & Hospital Environmental Control, 1985," NTIS PB85-923404, 1985.

4. Doucet, L.G., "Hospital/Infectious Waste Incineration Dilemmas & Resolutions," presented at the 1st National Symposium on Incineration of Infectious Wastes, Washington, D.C., May 6, 1988.

5. Doucet, L.G., "State-of-the-Art Hospital & Institutional Waste Incineration: Selection, Procurement and Operations," presented at the 75th Annual Meeting of The Association of Physical Plant Administrators of Universities and Colleges, Washington, D.C., July 24, 1988.

6. Perkins, J.J., Principles and Methods of Sterilization in Health Sciences, Published by Charles C. Thomas, 1983.

7. U.S.EPA, Guide For Infectious Waste Management, EPA 530-SW86-014, NTIS PB86-199130, May 1986.

VACCINES: HOW EFFECTIVE ARE THEY?

Adamadia Deforest

Department of Pediatrics
Temple University School of Medicine
St. Christopher's Hospital for Children
Philadelphia, Pennsylvania 19133

INTRODUCTION

In the United States, vaccines for human use are licensed by the Food and Drug Administration, and recommendations for their use are formulated and revised on a regular basis by the Advisory Committee for Immunization Practices (ACIP) of the Centers for Disease Control, and the Committee on Infectious Diseases of the American Academy of Pediatrics. Over the past twenty five years, immunization programs have markedly reduced the incidence of vaccine-preventable diseases in children, as evidenced by drastic declines in the rates of diphtheria, tetanus, pertussis, poliomyelitis, measles, mumps and rubella infections.

Unfortunately, successes already achieved in the pediatric population have not been realized in our adult population, prompting the government in late 1984 to issue recommendations for this group (2). The rationale for these new initiatives in adult immunization included the extraordinary successes of childhood immunization programs, the non-use or inappropriate use of vaccines such as diphtheria, tetanus, and influenza in adults, and the poor utilization of pneumococcal and hepatitis B vaccines in adults. Following the lead of the ACIP, the American College of Physicians in 1985 introduced a guide for adult immunization (19) which followed by two years an article containing guidelines for infection control in hospital personnel (47).

Over twenty vaccine preparations are currently available in the United States (9). Bacterial vaccines are licensed to prevent tuberculosis, cholera, diphtheria, tetanus, pertussis, Haemophilus influenzae type b, meningococci, plague, pneumococci, and typhoid fever. Virus vaccines are available to prevent hepatitis B, influenza, poliomyelitis, measles, mumps, rubella, rabies, and yellow fever. While some of these vaccines are recommended for all healthy infants and children (20), others are reserved for use in certain high-risk groups, for individuals in special occupations, or for travelers to areas of the world where the infectious agent is endemic or epidemic.

In the United States, approximately five million persons work in more than seven thousand hospitals. These individuals may become infected with vaccine-preventable bacteria and viruses through exposure to patients or by exposure outside the hospital setting. In either case, the infected health care personnel may in turn transmit infection to patients, other hospital personnel, members of their households, or other community contacts.

In this chapter, emphasis will be limited to those infection control objectives which should be a part of all hospital personnel health services for five vaccine-preventable viral infections: hepatitis B, influenza, measles, rubella, and poliomyelitis. Unfortunately, optimum utilization of these vaccines by health care workers has not yet been achieved. The chief impediments to vaccine usage have included public ignorance, professional apathy, the cost of vaccines, the cost of vaccine administration, a lack of insurance coverage, and lack of adequate record keeping on the part of physicians and nurses who staff employee health offices.

It is important to emphasize at the outset that currently available vaccines are indeed effective. But they are only effective if they are administered to the populations for which they were developed and licensed.

HEPATITIS B VACCINES

Utilization of hepatitis B vaccines has been hindered by cost as well as by unwarranted concerns over possible transmission of human immunodeficiency virus. The low levels of use of hepatitis B vaccine among health care workers are well documented (10,16,25,31,37,43).

The immunizing agent in the first hepatitis B vaccine (Heptavax-B®) which was licensed in 1982 consisted of purified inactivated hepatitis B surface antigen derived from human plasma. To date, over 4.5 million doses have been distributed in the United States. An estimated 1,400,000 persons have completed the three dose primary immunization series. Original vaccine recommendations focused primarily on three high risk groups: (1) health care workers, (2) staff and clients of institutions for the developmentally disabled, and (3) staff and patients in hemodialysis units. Although no precise figures are available, it is estimated that 85% of vaccine distributed between 1982 and 1987 was used for these three groups.

In 1986 a new genetically engineered vaccine (Recombivax-HB®) was licensed, providing an alternative to plasma-derived hepatitis B vaccine. This second generation vaccine is as immunogenic as the plasma-derived vaccine. When given in a three dose series, the recombinant vaccine induces protective antibodies in over 95% of healthy adult recipients. Major target populations for which the vaccine is recommended include male homosexuals, intravenous drug abusers, and health care personnel who are occupationally exposed to blood or blood-contaminated body fluids during the course of their work (4).

To the dismay of government agencies, acceptance of hepatitis B vaccine by health care personnel has been poor. Reasons for the poor acceptance rate include concerns regarding vaccine safety, particularly the frequency of side effects or fear of contracting acquired immunodeficiency syndrome, a need for more information about the vaccine, the high cost of the vaccine, questions regarding possible effects on present or future pregnancies, and individual ignorance regarding the risk of occupationally-acquired hepatitis B infection.

Development of vaccination programs for health care workers has progressed steadily since vaccine licensure (4). Several surveys of hospitals in 1985 showed that between 49% and 68% had established hepatitis B vaccination programs and that the number has increased steadily each year. Large hospitals (over 500 beds) were most likely to establish programs (90%). However, by June 1985, 60% of hospitals with fewer than 100 beds also had begun vaccination programs, and in 77%, the hospital paid for these vaccinations.

In spite of these programs, the actual use of vaccine in high risk health care professionals has been modest. In one statewide survey only 24% of high-risk personnel had received vaccine. In a survey of three large cities, only

24% of physicians had received vaccine. National surveys have shown higher rates of vaccine acceptance among dentists (44% by 1986) and hemodialysis staff (44% by 1985). However, even these rates fall well short of optimal coverage.

In a recent survey of physician acceptance at a university medical center (27), each of three attitudes:(1) that the vaccine is safe, (2) that the vaccine is effective, and (3) the perceived risk was high of contracting hepatitis B at work - was associated with a higher rate of vaccination. Overall, 79% of medical students, 61% of housestaff, and 35% of faculty indicated that they had received vaccine, for a total acceptance rate of 57%. Respondents who did not perceive themselves to be at high risk and who did not feel that the vaccine was safe or effective had the lowest vaccination rate, 29%. Those who felt they were at high risk and that the vaccine was both safe and effective had a vaccination rate of 79%. Oddly enough, medical students, the study group with the least amount of training and the only group who had to pay for the vaccine, had the highest vaccine acceptance rate.

Recommendations for immunization of hospital personnel with hepatitis B vaccine are as follows (4,48,49): (1) Persons with substantial risk of infection who are demonstrated or judged likely to be susceptible should be actively immunized; (2) Before immunizing, serologic screening for hepatitis B need not be done unless the hospital considers it cost effective or the employee requests it; (3) Prophylaxis with an immune globulin (passive immunization) should be used when indicated, such as following a needle-stick exposure to blood that is at high risk of being infectious; and (4) Immune globulin should not be used as a substitute for active immunization.

INFLUENZA VACCINE

It has been well established that influenza vaccine reduces death, hospitalization, and clinical illness in years when vaccine and epidemic strains of virus are similar. Influenza vaccine is up to 90% effective in preventing illness in young healthy adults but less effective in those over 65 years of age (49). Annual influenza vaccination has been considered the most important measure in preventing or attenuating influenza virus infections. Since 1963, the ACIP has recommended annual influenza vaccination of medical personnel caring for high-risk patients (5). Unfortunately, many health care professionals are reluctant to be vaccinated because of unfounded concerns about adverse reactions or doubts about vaccine efficacy (35).

Several studies have illustrated the potential for spread of influenza from health care workers to patients (11,23,36,45), and increases in the cost of health care due to employee absences caused by influenza infection (23,26). More definitive data concerning the occupational risk of influenza and the efficacy of influenza vaccine in reducing nosocomial spread are necessary to convince medical personnel of the need for annual vaccination (13,22,36,45).

The impact of influenza during years of epidemic virus activity cannot be overstated. Ten thousand or more excess deaths have been documented in each of 19 different epidemics during the years 1957-1986; more than 40,000 excess deaths occurred in each of several recent epidemics (8). Approximately 80-90% of the excess deaths attributed to pneumonia and influenza were among persons over 65 years of age. Because the number of elderly persons in the United States population is increasing, the toll from influenza can be expected to increase unless control measures are intensified. The number of younger persons at high risk for influenza-related complications is also increasing, and includes long-term survivors of organ transplantation, chronic lung diseases such as cystic fibrosis, and the increasing number of patients with human immunodeficiency virus (HIV) infection.

Vaccination of high-risk persons each year before the influenza season is underway is the single most important measure for reducing the impact of this disease. Health care workers can transmit infection to high-risk patients while the workers themselves are incubating infection, undergoing subclinical infection, or working despite feeling ill. Efforts to protect high-risk patients may be improved by reducing the chances that medical workers may expose them to infection. Therefore, the ACIP recommends that the following groups should be vaccinated on an annual basis: (1) Physicians, nurses, and other personnel who have extensive contact with high-risk patients (e.g. primary care and certain subspeciality clinicians and staff of chronic care facilities and intensive care units including neonatal intensive care units); and (2) Providers of home care to high-risk patients (e.g. visiting nurses and volunteer workers) as well as all household members of high-risk persons, including children, whether or not they provide direct care (8).

The importance of employee vaccination programs in preventing nosocomial influenza infections is illustrated by a recent prospective study in a Los Angeles acute care hospital (45). During the 1986-1987 season, a total of 43 cases of influenza A were identified. Seventeen cases (40%) occurred among working hospital employees. Fourteen cases were identified among patients in the emergency room or outpatient clinics, ten were community-acquired cases among hospitalized patients, and two cases were nosocomially acquired.

Another recent study describes an outbreak of influenza A among hospital personnel and patients in an Illinois acute care teaching hospital (36). Surveillance during the 1984-1985 season revealed influenza infection in 118 hospital personnel and 49 patients. The hospital staff was largely unvaccinated prior to the outbreak. Control of the outbreak was accomplished by an aggressive immunization program. Personnel were offered influenza vaccine at their work stations, at minimal inconvenience and with only minor disruption of their work activities. These efforts resulted in successful vaccination of approximately one third of the medical and nursing staff.

An aggressive immunization policy within hospitals, such as described above during the largest documented hospital outbreak to date, should increase compliance among medical personnel with current ACIP recommendations, and ensure protection for both staff and patients prior to the start of each influenza season.

MEASLES VACCINE

Since measles vaccine was licensed in 1963, the incidence of measles and its serious complications (encephalitis and death) has declined more than 98% compared to rates in the prevaccine era. Vaccination against measles has prevented approximately 52 million cases, 5200 deaths, and 17,400 cases of mental retardation (15). Despite this enviable achievement, outbreaks of measles continue to occur among adolescents and young adults in schools, universities, and the workplace. This potential for outbreaks exists because 5-20% of young adults remain susceptible to measles.

Spread of measles among adults in medical settings, including transmission involving medical personnel, is unacceptably common. A study conducted by the Centers for Disease Control summarized measles transmission in medical settings between 1980 and 1984 (21). Included were hospital wards, emergency departments, outpatient clinics, intensive care units, nonhospital emergency centers and outpatient clinics, physician's offices, laboratories, pharmacies, and drug rehabilitation centers. Excluded were other institutions that provide long-term care such as homes for the handicapped or mentally impaired. A total of 241 cases of measles were identified, representing 1.1% of all reported cases during this five year period. Sixty-four cases (35%) occurred in patients, including 21 who were hospital inpatients. Fifty-seven (24%) of the

241 cases occurred in medical personnel, including 23 nurses, 11 physicians, and five laboratory technologists. Another 17 nurses and six physicians were reported to have had measles illness, although these cases did not meet the case definition for acquiring measles in a medical setting. Thus, a total of 17 physicians and 40 nurses were infected during the period of study. Of 120 cases for whom the pattern of transmission was known, patient-to-patient spread (50%) and patient-to-staff spread (37%) were most common. Although personnel rarely transmitted disease to others, the authors conclude that more attention needs to be given to preventing the spread of measles in medical facilities, including assurance that medical personnel are immune to measles.

Adequate evidence of measles immunity includes (1) vaccination with live measles vaccine on or after the first birthday, (2) laboratory evidence of measles immunity, or (3) a history of physician-diagnosed measles. Experience has shown that mandatory immunization programs for medical personnel are in general more effective than voluntary programs (34).

A second study from the Centers for Disease Control described a measles outbreak in 1985 in a university medical setting in Michigan, which involved health care personnel (41). Nine students and five non-student contacts were affected. Eight of these cases were exposed to measles in medical settings, which included a general medicine ward, the student health center, and the hospital pharmacy.

While only a small proportion (2-3%) of all measles cases occur in medical personnel, as the above studies illustrate, the risk of infection in such persons is up to twelvefold higher than in the general population. Additionally, infected health care workers can also transmit measles to patients who may be at increased risk of complications due to underlying medical conditions.

For these reasons, the ACIP recommends that medical facilities ensure that all employees born after 1956 have proof of immunity (6). Since a substantial proportion of medical personnel who have acquired measles were born before 1957, the ACIP further states that medical facilities may also consider requiring proof of measles immunity for older employees who may have occupational exposure to measles.

RUBELLA VACCINE

Before rubella vaccines became available in 1969, most rubella cases occurred among school-aged children. By 1977, vaccination of children resulted in a marked decline in the reported incidence of rubella among children. However, the United States strategy of immunizing children had less effect on the reported incidence of rubella infection in persons 15 years of age or older (3). Approximately 10-20% of adolescents and young adults in the United States continue to be susceptible to measles, a proportion similar to that of the prevaccine era (39). Persons 15 years of age or older accounted for almost twice as many cases in 1982 as in 1981 (62% compared with 36%). The greatest increase in reported rates within this group occurred in those 25-29 years of age.

Recent outbreaks of rubella in hospitals (24,38,42,46) has led to the recommendation of compulsory rubella vaccination for both male and female hospital employees (2,3,19,24).

The following recommendations should be considered by hospitals in formulating rubella vaccine policy: (1) All personnel (male or female) who are considered to be at increased risk of contact with patients with rubella or who are likely to have direct contact with pregnant patients should be immune to

rubella; (2) Before immunizing, serologic screening for rubella is not
necessary unless the hospital considers it cost effective or the employee
requests it; (3) Persons can be considered susceptible unless thay have
laboratory evidence of immunity or documented immunization with live virus
vaccine on or after their first birthdat; (4) A clinical diagnosis of rubella
is unreliable and should not be considered in assessing immune status; and
(5) Consideration should be given to using rubella vaccine in combination
with measles and mumps (MMR) vaccine.

Since licensure of live attenuated rubella vaccine in 1969, the ACIP has
emphasized that pregnancy is a contraindication to rubella vaccination
because of concerns regarding the theoretical possibility of adverse effects
on the developing fetus. Data accumulated between 1971 and 1979 on 290 infants
born to 538 women who were inadvertently vaccinated within three months before
or after conception revealed that none of the infants had defects indicative
of the congenital rubella syndrome. This included 94 live-born infants of
women who were known to be susceptible before receiving the vaccine (12,18).

In 1979, the RA 27/3 rubella vaccine was licensed for use in the United
States. Concerns were raised that this new live attenuated virus vaccine might
have greater fetotropic and teratogenic potential than the earlier vaccines
because the vaccine virus was isolated from and propagated in human tissue.
Between 1979 and 1988, data accumulated on 212 infants born to 272 women known
to be susceptible before receiving the RA 27/3 vaccine revealed that none had
defects indicative of congenital rubella syndrome (18).

Although no congenital rubella syndrome-like defects have been noted in
either of the above long-term surveillance studies, it is known that rubella
vaccine viruses can cross the placenta and infect the fetus (40). Thus, because
of this evidence and because the theoretical risk to the fetus, however small,
cannot be absolutely ruled out, the ACIP continues to state: (1) Pregnancy
remains a contraindication to rubella vaccination; (2) Reasonable precautions
should be taken to preclude vaccination of pregnant women, including asking
women if they are pregnant, excluding those who say they are, and explaining
the theoretical risks to the others; and (3) If vaccination occurs within
three months before or after conception, the risk of congenital rubella syndrome is so small as to be negligible. Thus, inadvertent vaccination of a
pregnant woman should not be a reason in inself to consider interruption of
pregnancy (17).

POLIOVIRUS VACCINES

Two types of poliovirus vaccines are currently available in the United
States. Inactivated poliovirus vaccine (IPV) was licensed in 1955 and was used
widely until oral poliovirus vaccine (OPV) became available during tne period
1961-1964. Thereafter, the use of IPV rapidly declined to a level of less than
1% of all polio vaccine distributed annually in the United States. A primary
immunization series with either vaccine produces immunity to all three poliovirus serotypes in over 95% of recipients.

Poliomyelitis has become extremely rare in the United States since the
introduction of poliovirus vaccines. From a high of 21,269 cases of paralytic
polio in 1952, the incidence has declined to a low of only two cases in 1986
(29). Most recent cases of polio in this country have been associated with
ingestion of OPV or contact with vaccine excreted by an OPV recipient (30,33).

A method of producing a more potent IPV with greater antigenic content
was developed in 1978 and led to licensure of a new IPV which is produced in
human diploid cells (44). Several clinical trials with this new IPV have confirmed that seroconversion occurs in over 95% of vaccine recipients following

a full series of three doses, and a duration of antibody persistence of at least five years (14,32).

Although both IPV and OPV are effective in preventing poliomyelitis, OPV is the vaccine of choice for primary immunization of children in the United States (1). OPV is preferred because it induces intestinal immunity, is easy to administer, is well accepted by parents since no injection is required, and has successfully eliminated disease associated with wild polioviruses in the United States.

The ACIP further states that routine primary poliovirus vaccination of adults (generally those 18 years or older) residing in the United States is not necessary, since most are already immune and there is only a very small risk of exposure to wild poliovirus strains. Immunization is recommended for certain adults who are at greater risk of exposure to wild polioviruses than the general population, including: (1) Travelers to areas or countries where poliomyelitis is endemic or epidemic; (2) Members of communities or specific population groups with disease caused by wild polioviruses; (3) Laboratory workers handling specimens which may contain polioviruses; and (4) Health-care workers in close contact with patients who may be excreting polioviruses.

For unvaccinated adults at increased risk of exposure, a primary series of enhanced potency IPV is recommended (7). IPV is preferred because the risk of vaccine-associated paralysis following OPV is slightly higher in adults than in children and because health care personnel may shed virus after OPV and inadvertently expose susceptible or immunocompromised patients to live virus.

Adults who are at increased risk of exposure and have had (1) at least one dose of OPV, (2) fewer than three doses of conventional IPV, or (3) a combination of conventional IPV and OPV totalling fewer than three doses should receive at least one dose of OPV or enhanced potency IPV. Additional doses needed to complete a primary series should be given if possible.

Adults who are at increased risk of exposure and who have previously completed a primary series with any one or a combination of polio vaccines can be given one dose of eith OPV or enhanced potency IPV.

SUMMARY

It is the responsibility of all health care workers to promote the concepts of adult immunization practices. Pediatricians have accepted for many years their obligation to protect infants and children from vaccine-preventable diseases. By now we should have learned from their example. Because an increased proportion of vaccine-preventable diseases now occur in adults, and because health care workers are at higher risk than the general population to infection with certain agents including those discussed above, it is time that all of us implement the recommendations outlined here. We owe it to ourselves, and we certainly owe it to the patients we care for.

LITERATURE CITED

1. Advisory Committee on Immunization Practices. 1982. Poliomyelitis prevention. MMWR 31:22-26,31-34.
2. Advisory Committee on Immunization Practices. 1984. Adult immunization. MMWR 33(suppl):1S-68S.
3. Advisory Committee on Immunization Practices. 1984. Rubella prevention. MMWR 33:301-310,315-318.
4. Advisory Committee on Immunization Practices. 1987. Update on hepatitis B prevention. MMWR 36:353-360,366.

5. Advisory Committee on Immunization Practices. 1987. Prevention and control of influenza. MMWR 36:373-380,385-387.
6. Advisory Committee on Immunization Practices. 1987. Measles prevention. MMWR 36:409-418,423-425.
7. Advisory Committee on Immunization Practices. 1987. Poliomyelitis prevention: Enhanced-potency inactivated poliomyelitis vaccine - Supplementary statement. MMWR 36:795-798.
8. Advisory Committee on Immunization Practices. 1988. Prevention and control of influenza. MMWR 37:361-364,369-373.
9. Advisory Committee on Immunization Practices. 1989. General recommendations on immunization. MMWR 38:205-214,219-227.
10. Anderson, A.C., and G.R. Hodges. 1983. Acceptance of hepatitis B vaccine among high-risk health care workers. Am. J. Infect. Control 11:207-211.
11. Atkinson, W.L., N.H. Arden, P.P. Patriarca, N. Leslie, K.J. Lui, and R. Gohd. 1986. Amantadine prophylaxis during an institutional outbreak of type A (H1N1) influenza. Arch. Intern. Med. 146:1751-1756.
12. Bart, S.W., H.C. Stetler, and S.R. Preblud. 1985. Fetal risk associated with rubella vaccine: An update. Rev. Infect. Dis. 7(suppl 1):S95-S102.
13. Berlinberg, C.D., S.R. Weingarten, L.B. Bolton, and S.H. Waterman. 1989. Occupational exposure to influenza: Introduction of an index case to a hospital. Infect. Control Hosp. Epidemiol. 10:70-73.
14. Bernier, R.H. 1986. Improved inactivated poliovirus vaccine: An update. Pediatr. Infect. Dis. 5:289-292.
15. Bloch, A.B., W.A. Orenstein, and H.C. Stetler. 1985. Health impact of measles vaccination in the United States. Pediatrics 76:524-532.
16. Bodenheimer, H.C., J.P. Fulton, and P.D. Kramer. 1986. Acceptance of hepatitis B vaccine among hospital workers. Am. J. Public Health 76:252-255.
17. Centers for Disease Control. 1987. Rubella vaccination during pregnancy - United States, 1971-1986. MMWR 36:457-461.
18. Centers for Disease Control. 1989. Rubella vaccination during pregnancy - 1971-1988. MMWR 38:289-293.
19. Committee on Immunization, Council of Medical Societies. 1985. Guide for adult immunization. American College of Physicians, Philadelphia.
20. Committee on Infectious Diseases, American Academy of Pediatrics. 1988. Report of the Committee on Infectious Diseases, ed. 21. American Academy of Pediatrics, Elk Grove Village, Illinois.
21. Davis, R.M., W.A. Orenstein, J.A. Frank, J.J. Sacks, L.G. Dales, S.R. Preblud, K.J. Bart, N.M. Williams, and A.R. Hinman. 1986. Transmission of measles in medical settings: 1980 through 1984. J.A.M.A. 255:1295-1298.
22. Fedson, D.S. and H.A. Kessler. 1983. A hospital-based influenza immunization program, 1977-1978. Am. J. Public Health 73:442-445.
23. Fralick, R.A. 1985. Absenteeism among hospital staff during an influenza epidemic. Can. Med. Assoc. J. 133:641-642.
24. Greaves, W.L., W.A. Orenstein, H.C. Stetler, and S.R. Preblud. 1982. Prevention of rubella transmission in medical facilities. J.A.M.A. 242:861-864.
25. Haas, R., and M.E. Beideman. 1986. Marketing a hepatitis B vaccine program. Infect. Control 7:339-341.
26. Hammond, G.W. and M. Cheang. 1984. Absenteeism among hospital staff during an influenza epidemic: Implications for immunoprophylaxis. Can. Med. Assoc. J. 131:449-452.
27. Hashimoto, F., W.C. Hunt, and P. Brusuelas. 1988. Physician acceptance of the hepatitis B vaccine at a university medical center. Am. J. Public Health 78:973-974.
28. Hoffman, P.C., and R.E. Rixon. 1977. Control of influenza in the hospital. Ann. Intern. Med. 87:725-728.
29. Katz, S.L. 1987. Controversies in immunization. Pediatr. Infect. Dis. J. 6:607-613.
30. Kim-Farley, R.J., K.J. Bart, and L.B. Schonberger. 1984. Poliomyelitis in the U.S.A.: Virtual elimination of disease caused by wild virus. Lancet 2:1316-1317.

31. Klimek, J.J., L. Brettman, E. Neuhaus, and R.A. Garibaldi. 1985. A multi-hospital hepatitis B vaccine program: Prevalence of antibody and acceptance of vaccination among high-risk hospital employees. Infect. Control 6:32-34.

32. McBean, A.M., M.L. Thoms, and R.H. Johnson. 1984. A comparison of the serologic responses to oral and injectable trivalent poliovirus vaccines. Rev. Infect. Dis. 6(suppl 2):S552-S555.

33. Nkowane, B.M., S.G.F. wassilak, and W.A. Orenstein. 1987. Vaccine-associated paralytic poliomyelitis - United States: 1973 through 1984. J.A.M.A. 257:1335-1340.

34. Orenstein, W.A., P.N.R. Heseltine, and S.J. LeGagnoux. 1981. Rubella vaccine and susceptible hospital employees: Poor physician participation. J.A.M.A. 245:711-713.

35. Pachucki, C.T., J.R. Lentino, and G.G. Jackson. 1985. Attitudes and behavior of health care personnel regarding the use and efficacy of influenza vaccine. J. Infect. Dis. 151:1170-1171.

36. Pachucki, C.T., S.A. Walsh-Pappas, G.F. Fuller, S.L. Krause, J.R. Lentino and D.M. Schaaff. 1989. Influenza A among hospital personnel and patients: Implications for recognition, prevention, and control. Arch. Intern. Med. 149:77-80.

37. Palmer, D.L., and R. King. 1983. Attitude toward hepatitis vaccination among high-risk hospital employees. J. Infect. Dis. 147:1120-1121.

38. Polk, B.F., J.A. White, P.C. DeGirolami, and J.F. Modlin. 1980. An outbreak of rubella among hospital personnel. New Engl. J. Med. 304:541-545.

39. Preblud, S.R., M.K. Serdula, J.A. Frank, A.D. Brandling-Bennett, and A.R. Hinman. 1980. Rubella vaccination in the United States: A 10-year review. Epidemiol. Rev. 2:171-194.

40. Preblud, S.R., H.C. Stetler, J.A. Frank, W.L. Breaves, A.R. Hinman, and K.L. Herrmann. 1981. Fetal risk associated with rubella vaccine. J.A.M.A. 246:1413-1417.

41. Sienko, D.G., C. Friedman, H.B. McGee, M.J. Allen, W.F. Simonsen, B.B. Wentworth, T.C. Shope, and W.A. Orenstein. 1987. A measles outbreak at university medical settings involving health care providers. Am. J. Public Health 77:1222-1224.

42. Strassburg, M.A., D.T. Imagawa, and S.L. Fannin. 1981. Rubella outbreak among hospital employees. Obstet. Gynecol. 57:283-288/

43. Tong, M.J., A.M. Howard, G.C. Schatz, M.A. Kane, D.A. Roskamp, R.L. Co, and C. Boone. 1987. A hepatitis B vaccination program in a community teaching hospital. Infect. Control 8:102-107.

44. von Seefried, A., J.H. Chus, J.A. Grant, L. Letvenuk, and E.W. Pearson. 1984. Inactivated poliovirus vaccine and test development at Connaught Laboratories Ltd. Rev. Infect. Dis. 6(suppl 2): S345-S349.

45. Weingarten, S., M. Friedlander, D. Rascon, M. Ault, M. Morgan, and R.D. Meyer. 1988. Influenza surveillance in an acute care hospital. Arch. Intern. Med. 148:113-116.

46. Weiss, K.E., C.E. Falvo, E. Buimovici-Klein, J.W. Magill, and L.Z. Cooper. 1979. Evaluation of an employee health service as a setting for a rubella screening and immunization program. Am. J. Public Health 69:281-283.

47. Williams, W.W. 1983. Guideline for infection control in hospital personnel. Infect. Control 4:326-349.

48. Williams, W.W. and J.S. Gardner. 1986. Personnel health services. In Hospital Infections, 2nd ed., J.V. Bennett and P.S. Brachman, eds., Little, Brown and Company, Boston.

49. Williams, W.W., M.A. hickson, M.A. Kane, A.P. Kendal, J.S. Spika, and A.R. Hinman. 1988. Immunization policies and vaccine coverage among adults: The risk for missed opportunities. Ann. Intern. Med. 108:616-625.

LEGAL ASPECTS OF INFECTION CONTROL

Ellen Covner Weiss

Hospital of the University of Pennsylvania
3400 Spruce Street
Philadelphia, PA 19104

I. Introduction

In this discussion[1] of Legal Aspects of infection
control, Infection Control refers to a hospital program
that is designed to prevent, identify and control
infections acquired in the hospital or brought into the
hospital from the community. This program is pervasive and
comprehensive. It affects patients, employees,
professional staff, treatment programs, and hospital
operations, for example the disposal of hazardous waste or
sterilization of equipment and devices. Indeed, the
comprehensive nature of a hospital infection control
program may be likened to public health programs undertaken
in the community. The hospital is in effect a
mini-community. Concerns about containing and treating
venereal disease or AIDS and the management of infectious
waste are all concerns reflected at a broader level in the
community and in the state through authorizing legislation
and a variety of programs.

II. Authority for Infection Control

The legal authority for an infection control program
rests primarily with state and local legislation. On a
broad level there is usually state legislation which
establishes the goal of prevention and treatment of
communicable and other diseases for the protection of the
public and the promotion of health. The legislation
usually authorizes a variety of infection control
activities. Subsequent legislation may specifically mandate
certain actions directed toward a particular health
problem. Further, state hospital licensure laws and
regulations mandate specific infection control programs and
activities. At the local level, municipal ordinances and
regulations may be designed to implement or add further
definition to the authorizing state legislation and
regulations.

Turning first to state authorization, the Pennsylvania Disease and Prevention Control Law of 1955[2] is a good example of how these statutes work. This statute replaced prior public health statutes. Its focus is on communicable diseases and it makes local boards and departments of health primarily responsible for implementing this state-wide mandate. It also devotes a lot of attention to schools as a source of communicable disease and emphasizes the need to prevent the spread of disease in schools as well as in health care facilities.

In addition to delegating authority to local boards and departments of health, the state statute imposes direct reporting requirements on a number of individuals. For example, physicians are required to report any person suffering from or "suspected" of having a communicable disease or being a carrier.[3] This reporting requirement also extends to the heads of hospitals and other institutions, directors of laboratories, school authorities, proprietors of hotels, roentgenologist, proprietors of boarding rooms or lodging houses, nurses, midwives, householders and other persons having knowledge or suspecting someone of carrying a communicable disease.[4] This statute also requires physicians and others in charge of health institutions to report other (undefined) diseases and conditions that are needed to protect and promote the health of people and prevent the spread of disease.[5] Thus the effect is to go beyond communicable diseases as a public health concern and to look towards other diseases or problems that may endanger the public health.

The statute provides for isolation, quarantine or other control measures needed to control the spread of communicable diseases.[6]

The statute also provides that physical examinations may be compelled for persons suspected of carrying or having venereal disease, tuberculosis or other communicable diseases.[7]

The statute authorizes that the individual be quarantined until it is determined whether he/she has a communicable disease or, alternatively, allows the appropriate authority to file a petition in the Pennsylvania Court of Common Pleas before a judge without a jury to request that an examination be ordered if the individual has "no valid reason for the refusal".[8] The judge may order that the individual be committed to a treatment institution.[9] It is important to note that, in the absence of an examination to determine that the individual does have a communicable disease, to commit a possibly healthy person to a treatment institution where he/she may be exposed to diseases is somewhat inconsistent.

The statute devotes considerable attention to venereal disease. An individual convicted of a sex offense or in custody or charged with a sex offense may be compelled to have a venereal disease examination and to receive treatment.[10] The statute also requires prenatal

examinations for syphilis.[11] The treating physician is
to take a blood test on the pregnant women "unless the
women dissents".[12] In the event of dissent, it is the
duty of the physician to explain to the pregnant woman the
desirability of such a test.[13] Although this requirement
appears separately from other Pennsylvania statutory
provisions involving informed consent, it seems clear that
the most appropriate way for the physician to test for
syphilis is to start first with an explanation to the
patient of the purpose of the blood test, rather than
drawing blood unless the woman dissents. Indeed, unless
the physician tells the patient why he/she wants to take
the blood, the patient would be unlikely to dissent or
protest because she would have no reason to suspect that
the blood was being taken for public health purposes such
as testing for syphilis, with the potential for
stigmatization or other kinds of complications. On the
other hand, in the years since the statute was passed, it
may be that acceptance of the epidemic nature of syphilis
is so widespread that patients have come to expect such
testing as a routine matter.

The physician's responsibility does not end with
testing the pregnant woman for syphilis. In fact, when
reporting a birth or a fetal death, the physician is also
required to state whether the syphilis test occurred and if
not, why not.[14]

To further underscore the concern about containing the
spread of venereal disease, the statute provides that
patients under twenty-one may be treated for venereal
disease without parental consent.[15] This provision
reflects the legislative judgment that minors may be
deterred from seeking needed medical treatment if their
parents were aware of the nature of the illness. In
balancing parental control over minors against the health
needs of those minors and the risk to others from contact
with untreated minors, health needs prevail. If the minor
consents, the physician is immunized from liability for
non-negligent treatment.[16] This means that the parents
of the minor who was being treated for venereal disease
could not successfully claim that treating the minor
without their consent was negligence or assault and
battery. Of course the statute does not immunize the
physician from liability for negligent medical treatment
and, in that event, presumably the parents as well as the
patient could sue for the consequences of the negligent
treatment.[17]

Finally, the statute provides that data be kept
confidential except for research or when necessary to carry
out the purposes of the act.[18]

The Pennsylvania statute is typical of those state
statutes which provide very broad overall authority, with
delegation to state and local boards to implement most of
the details of the act. In considering its date of
passage, 1955, its emphasis on venereal disease as the
communicable disease of the moment is understandable.
However, it is interesting to observe that authorization of

quarantine, compelled examination, or testing of individuals charged with sex offenses are all part of the debate today over the prevention and treatment of AIDS.

State licensure boards may underscore the physician's role in infection control through licensure regulations. There may be a general statement about practicing in accordance with accepted medical standards. In addition, there may be a more specific statement. For example, a regulation may require physicians not to refuse to treat patients who have AIDS, solely because of their disease, or else risk loss of licensure.

Other statutes parallel the provisions of the Pennsylvania Disease and Prevention Control law. In addition some provide incentives for specific infection control measures. For example, California immunizes "a person" for any injury (including residual effects) caused by vaccinating minors as required by state law, except for willful misconduct or gross negligence on the part of the vaccinator.[19] In the 1986 case, <u>Flood v. Wyeth Laboratories, Inc.</u>,[20] "a person" was defined to exclude a manufacturer. The court rejected Wyeth's argument that statutory immunization was necessary to assure a supply of the DPT vaccine. It reviewed the legislative history of the provision and found that the intent to provide immunity was limited to physicians. The court said that a manufacturer could protect itself from the costs of litigation by increasing its prices. Should manufacturers cease to produce the vaccine because of litigation, the legislature could then provide them with immunity. Florida provides immunity to members of school boards and their employees, public and private.[21]

Another source of authority for infection control programs in hospitals is the Joint Commission on Accreditation of Health Care Organizations ("Joint Commission"). This is a private group that accredits hospitals. It is generally considered as advantageous for hospitals to be accredited by the Joint Commission since accreditation is deemed by the federal government to constitute compliance with the Medicare and Medicaid Conditions of Participation. The Joint Commission reviews hospitals in three-year cycles. Some states which follow the same three-year review period have a coordinated review with the Joint Commission and this relieves some of the review activity on a hospital in preparing to meet the requirements of these different bodies. The Joint Commission issues an accreditation manual for hospitals. It sets out in some detail requirements of an infection control program which covers the same goals as reflected in state legislation.[22] Namely, there is concern about containing the spread of communicable diseases along with their detection and prevention.

An additional source and kind of "authority" for infection control programs may arise from reports of various groups. The AIDS crisis has resulted in a number of such statements. These statements may review the medicine or law of AIDS and describe appropriate prevention, educational, or legal responses to the problem.[23] These reports are not legally binding. However, they may serve to set a standard of medical practice or behavior as defined by the profession itself. This standard may be referred to for purposes of malpractice or other kinds of litigation. The plaintiff may use the statement to try to establish the profession's definition of appropriate behavior which was violated by the defendant, thereby causing harm to the plaintiff.

Finally, judicial action is a source of legal authority that may incorporate both legislation and statements of private groups. Judicial action takes place at the federal and state levels. Since each state court system is different and is called upon to interpret state laws which differ from state to state, there may be a variety of actions and results taken. Courts frequently look to the decisions and reasoning of other courts when asked to decide a new issue. _Darling v. Charleston Community Memorial Hospital_[24] illustrates how a court may examine relevant decisions of other jurisdictions, as well as make use of privately developed standards. In this case, a father brought a malpractice action on behalf of his son. He alleged that negligent treatment caused a below-the-knee amputation of his son's leg and he sued the hospital and the physician. The hospital sought to defend itself on several grounds. It argued that it did not deliver health care itself, should not be found responsible for the acts of physicians which used the hospital facilities, if it used reasonable care in their selection, and should not be held responsible for the acts of its employee nurses which occurred while the nurse was executing the physician's orders. The court rejected the hospital's position. It found that there was an expectation of the public that the hospital had a responsibility to treat patients. The court referenced the Joint Commission hospital standards, state licensing regulations, and the hospital's bylaws to demonstrate the recognition and assumption of responsibility by the hospital to patients.

III. _Balancing Individual Rights and Infection Control Needs_

As discussed above, performance of an infection control program takes place in a hospital. The hospital has obligations to patients as well as to its employees and other staff, such as physicians. These obligations may, at times, compete or conflict with one another.

A.   _Health Care Workers_

As a general principle, physicians are not obligated to accept anyone and everyone as a patient. Those they accept should be treated within their area of expertise. There are several exceptions to this principle. State law or licensure board regulations may require a physician to render emergency medical treatment.[25] State law or licensure board regulations and/or federal law may prohibit refusals to treat patients because they are indigent[26] or under the recent anti-dumping provisions of the Social Security Act,[27] or where refusal is based on an individual's sex, race, national origin, sexual orientation, or real or perceived handicap.[29]

Once a physician has assumed treatment of a patient, he/she may be subject to some of the state reporting or other requirements discussed above. The physician may have treatment obligations imposed through the medical staff bylaws. For example, the bylaws or rules and regulations may describe the components of a patient's history and physical examination, who may do it and the time period after admission for it to occur.

A physician or other practitioner hired or appointed to direct or participate in the hospital's infection control program may have specific responsibilities for such a program. The Joint Commission's requirements for an infection control program include involvement of the staff in surveillance and reporting programs, policy development, and guidelines for personnel.[30] Staff not directly involved in the design or oversight of an infection control program may have treatment responsibilities which are developed by such a program, e.g. antibiotic prophylaxis for patients with certain illnesses or who are to undergo certain invasive procedures. The Center for Disease Control or public health department may recommend infection control measures.[31] Their recommendations in the prevention and treatment of AIDS, including the adoption of universal barrier precautions, have been incorporated by OSHA into its compliance procedures.[32] Here, the obligation to treat or act would be to use the universal barrier precautions and follow the recommendations of those programs. And finally, professional groups such as the American Medical Association[33] ("AMA") have said that physicians should not refuse to treat patients simply because the patient has AIDS and, in fact physicians have an obligation to treat these patients.

Looking at these different kinds of obligations and how they arise, one could say that statutory obligations to treat reflect a concern for promoting public well-being and require the physician or other health care worker to act to further that goal. Statements by the AMA and other professional groups emphasize that the physician's or other health care worker's obligation to treat is an integral part of his/her professional responsibility.[34]

Nonetheless there is recognition that in certain circumstances the health care worker may be relieved from the obligation to treat. One example of this is the "conscience clause" found in the federal regulations governing sterilizations and abortions.[35] This clause allows institutions that receive federal money to refuse to do sterilizations or abortions if those procedures are against the religious beliefs or moral conviction of the institution. The provision further exempts individuals, based on their religious beliefs or moral convictions, from performing or assisting in a sterilization or abortion.

There are many state statutes that relieve the physician or nurse from participating in an abortion.[36] Generally the statue also forbids discrimination against a person who refuses to participate in an abortion. The statute usually covers hospital employees, and some cover medical students or other health care students. Some require the objection to be on moral, ethical or religious grounds. A few also exempt staff from participating in sterilization or euthanasia. For example, in Illinois, the statute protects "the right of conscience of all persons who refuse to obtain, receive, or accept, or who are engaged in the delivery of medical services and medical care."[37] A few statutes prohibit discrimination against individuals who favor or will participate in an abortion.[38]

Another source of authority that has been alleged to justify a refusal to act has been the professional standards of a specialty group. For example in the 1985 New Jersey case, Warthen v. Toms River Community Memorial Hospital,[39] a nurse refused to continue dialyzing a terminally ill patient after observing him in cardiac arrest during dialysis. The nurse cited the American Nurse Association ("ANA") code for nurses, specifically the provision that states:

> "The nurse provides services with respect to human dignity and the uniqueness of the client unrestricted by considerations of social or economic status, personal attributes, or the nature of health problems."[40]

The interpretive statement allowed for a nurse to refuse to participate where she was personally opposed to treatment in a case because of the nature of the health problem or the procedure.

The nurse was dismissed after she had refused to dialyze the patient and her dismissal was upheld. The court found that the ANA statement was not one of public policy that the nurse could rely upon to justify her refusal to treat. Rather, it established a standard that was personally beneficial to nurses, but was not in the best interests of the public.

In the absence of statutory or regulatory relief from the obligation of a staff member to treat, there may be "conscience clause" type approaches in the medical staff bylaws or other hospital or medical staff policies. For example, many hospitals have policies on withholding or withdrawing treatment.[41] This policy may recognize that the withholding or withdrawal of treatment may conflict with the religious and/or ethical principles of staff. Accordingly, the health care worker may be able to transfer his/her responsibilities to the patient to another health care worker. This allows the hospital to meet the needs and rights of the patient while recognizing other needs of its staff.

Even without a formal policy, it usually is accepted practice to arrange for a substitute physician or for one nurse to "cover" for another. However, where there are few staff trained in a particular specialty, or a shortage of available staff, such as nurses, arranging for substitute coverage may be difficult.[42]

The job descriptions of health care workers, especially nurses, medical technologists, lab workers and others, serve as direction and authority for the individual's responsibilities. The job description may describe in great detail the activities that the individual is expected to undertake. In some hospitals, the employment application form explicitly states that the applicant may come into contact with individuals carrying infectious or other hazardous diseases so that individuals who apply will be aware that hospital employment may have health risks.

Using the job description as a baseline, it may be a violation of the hospital's personnel policies for an individual to refuse to treat or act in accordance with the job description. Examples would be a pregnant phlebotomist who refuses to draw blood on a high risk patient because she is afraid of contracting AIDS, or a housekeeper who refuses to clean a room of a patient having a communicable disease.

The requirements of job descriptions may be read in the context of the OSHA obligation that employers provide a safe work environment.[43] By their very nature, hospitals can never be totally risk-free. However, the OSHA obligation presumably is met if the hospital follows the CDC universal barrier precautions and other relevant regulations. The hospital's compliance with these measures can be understood as reducing the risk to the employee even if it is not completely eliminated.

This situation was addressed in the recent case of Stepp v. Review Board of the Indiana Employment Security Division.[44] In this case the court upheld the determination of the review board that a lab worker had been properly discharged. The worker had refused to perform laboratory test on vials of bodily fluids that had an AIDS warning.

The laboratory had developed safety manuals on handling these kinds of bodily fluids. The employee had been told that her refusal to handle these vials would result in her termination. In fact, prior to termination she had been suspended, was returned to work, and upon again refusing to handle the vials, was discharged. The employee said that she refused to handle the fluids because it would "violate God's will".

The court found that measures had been taken to protect lab workers, that the laboratory was a safe place to work, and that the worker was not justified in her refusal. The court also found that her reliance on the

OSHA requirement that the employer provide a safe work environment was misplaced.[45] The court examined a specific OSHA provision which prohibits the employer from discharging an employee who exercises a protected right. Here the applicable right would be the refusal to work because of a reasonable apprehension of death or serious injury and a reasonable belief that no less drastic alternative is available. In applying that standard to the facts of this case the court found that the worker's refusal was based on a religious, not a safety reason, which reason was not protected by OSHA. The effect of the court's decision was to deny the lab worker unemployment compensation.

Some employees might feel that the hospital environment is not safe, not because of their contact with patients or hazardous substances, but because they may be exposed to diseases carried by other employees. In particular, staff are concerned about exposure to AIDS. The hospital is not free to respond to their fears simply by discharging employees with AIDS or who are suspected of having the disease. Los Angeles[46] and Philadelphia[47] among others, have enacted ordinances prohibiting discrimination against employees with AIDS. Recent cases[48] have treated this situation as falling under the protection of Section 503 of the Rehabilitation Act of 1973.[49] Employees may not be discharged if they have AIDS as long as they may do their jobs with reasonable precautions for the safety of others. Remedies that are available include sending the individual home on a medical leave, transferring him/her to another position consistent with the person's work abilities which reduces the possibility of exposure, and so on. The Pennsylvania Medical Society[50] stresses the importance of educating employees in understanding the illness and how to prevent its transmission.

If a health care worker contracts a disease or infection while employed in a hospital, he/she might claim that the illness was work-related, thereby entitling the employee to relief under the workers' compensation statute.[51] The typical workers' compensation statute provides that, in exchange for the receipt of workers' compensation, the individual gives up the right to a negligence or other kind of action against the employer. However, the injury or illness must be found to be work related and this raises a question of causation. Specifically, did the employee's own behavior or negligence cause the injury or illness?

An example of this situation is a health care worker who contracts hepatitis B. Cases have gone both ways on the element of causation and whether or not workers' compensation could be provided. For example, in Sperling v. Industrial Commission,[52] an Illinois case, the plaintiff nurse alleged that she contracted hepatitis B in the course of her employment at the hospital. The plaintiff worked in the operating room and was in frequent contact with surgical instruments which, of course, had human blood on them. In addition, the nurse frequently

sustained minor cuts and other exposure to blood. She was
diagnosed as having hepatitis B after one year's employment
at the hospital. The plaintiff had to show that her
employment caused the disease. She had expert witnesses
testify about causation. They provided general information
and studies on health care workers, but they did not
testify about causation specifically as to operating room
nurses. The nurse also was unable to state the exact date
she contracted the infection, nor could she provide
evidence of the presence or absence of hepatitis B in the
patients to whom she had been exposed. The court decided
that direct proof of causation was not required. Instead,
the court accepted a weaker evidentiary standard and
permitted indirect proof. Namely, it allowed the plaintiff
to show that it was more likely than not that the cause of
her hepatitis B was her employment at the hospital. The
plaintiff offered evidence that she did not get the disease
through sexual activity or intravenous drug use. The
nature of the transmission of the disease made other
non-employment activities an unlikely cause of the
disease. Thus the court allowed the nurse to receive
workers' compensation.

        In contrast, in Georgia in the case of
Fulton-DeKalb Hospital Authority v. Bishop,[53] an
emergency medical technician claimed that his work placed
him at a greater risk for hepatitis B than if he were a
member of the general public. The court found that the
nature of his employment was not enough, by itself, to
establish a causal connection between his employment and
contracting hepatitis B. Accordingly he was denied
workers' compensation.

B.   Patients

        In looking at the rights and responsibilities of
patients and the operation of an infection control program
one must refer first to already established principles
regarding patients' rights. While the exact statement of
these rights varies from state to state, there is general
recognition of the patient's right to participate in
his/her health care, including consent or refusing consent
to recommended treatment. The Joint Commission has issued
a statement addressing patients rights and
responsibilities[54].

        The Pennsylvania Patient's Bill of Rights is
typical[55] of state statutes. It describes a number of
"minimal provisions" which may be supplemented by
hospitals. Nine of the twenty-two sections reference the
patient's right to consent or right of access to
information about his/her health care.[56] One states the
right of the patient "to know what hospital rules and
regulations apply to his conduct as a patient."[57]

        Other state statutes address consent in the
context of specific conditions. Examples include blood
tests for syphilis in pregnant women and minors' consent
for treatment for venereal disease, discussed above in
section II.

In addition, there is extensive case law throughout the country discussing the patient's right of informed consent and the applicable standard of disclosure (what is customary for physicians to tell patients _versus_ what the patient would deem material to his/her consent).

The focus of the statutes and judicial decisions is on the right of the patient to make decisions regarding his/her treatment that conform to the patient's views of what best meets his/her welfare. (This may go beyond a strictly medical or health-based view.)

This singular focus must be examined in the context of an infection control program where interests beyond that of the welfare of the individual are also important. The Joint Commission's statement about patients' rights and responsibilities includes the following:

Personal Safety

The patient has the right to expect reasonable safety insofar as the hospital practices and environment are concerned.[58]

While this provision appears to address aspects of physical safety, such as building code, housekeeping, and other matters, it has relevance for a variety of infection control concerns. Disposal of infectious waste, limiting contact of health care workers with communicable diseases, cleanliness of equipment, patient rooms and other items or areas with which patients have contact, all bear on the patient's safety. However, there is no discussion by the Joint Commission, either in this statement or in the standards for infection control[59] which explicitly addresses the need for patients to participate in infection control programs which may go beyond their individual treatment needs or conflict with their personal preferences. The Pennsylvania Bill of Rights[60] similarly does not clearly relate infection control needs to patient decision-making. One provision may be read to offer limited support for connecting them. It states:

A patient has the right to refuse any drugs, treatment, or procedure offered by the hospital, to the extent permitted by law, and a physician shall inform the patient of the medical consequences of the patient's refusal of any drugs, treatment, or procedure.[61] (Emphasis added).

This paragraph appears to allow for situations where the patient's right to refuse may be overridden. These situations are not defined in the Patient's Bill of Rights. Examples may include a court order for a patient to receive treatment he/she objected to, or administration of a drug or procedure expressly required by law, such as ocular prophylaxis for newborns or vaccination. However, it could be a basis for asserting that a patient should participate in an infection control program that goes beyond concern for the benefit to the individual to reflect a broader infection control concern. An example would be placing a patient with tuberculosis or other airborne infection in isolation. This offers no particular benefit to the patient but serves to reduce the risk that the disease will be spread to others. And it allows the hospital to meet state requirements for an infection control program.[62] Another example would be to require patients who are colonized with methicillin-resistant staphyloccocus aureus, but who are not symptomatic, to undergo a course of antibiotics prior to accepting them in a nursing home. This would protect other patients from exposure to the disease. However, antibiotics are not without side effects and may not be advisable for a particular patient. Thus there may be a question of protecting the health of others at the risk of some harm to a patient.

Balancing individual and societal concerns have been recognized in some instances by statute, frequently in the context of asserted religious beliefs. For example Delaware,[63] Massachusetts,[64] and Illinois[65] all have provisions which exempt school children from mandatory vaccination based on religious beliefs of the parents or legal guardian. In exempting the children from vaccination these statutes, in effect, make the determination that recognition of the religious belief outweighs concerns for the health of the particular child as well as the welfare of the other children. It is interesting to note, that it is not the child and his/her expression of religious beliefs that is being recognized but that of the religious beliefs of the parents. This is consistent with other case law and principles that recognize the parents as the natural guardian of the child and defer initially to the parents in determining what is in the child's best interests.

In contrast, there are mandatory reporting statutes which exist in some form in all states. These statutes may require reporting of venereal disease, AIDS, and other communicable diseases. The provisions may or may not permit reporting to be anonymous. In some instances identified reporting allows for a public health function to occur. Namely, the affected individual and others with whom he/she has come into contact are notified by the public health department to provide early diagnosis or disease prevention. These statutes reflect a determination that the individual's right of privacy is outweighed by the health needs of the community.

Individuals have protested public disclosure of information given in confidence to their physicians. Two cases illustrate this situation. In Horne v. Patton[66] the patient sued his physician. He asserted that disclosure of information about him violated the right of privacy and the fiduciary duty and contractual obligation that the physician owed him. The patient alleged that he had seen the physician for treatment, requesting that the physician not disclose any information. However, the

200

patient asserted that the physician gave confidential information about the patient to the patient's employer which caused the employer to fire him. Since there was no physician-patient privilege statute in Alabama the court looked to other relevant references to answer the claims raised by the patient.

The court first found that the Hippocratic Oath and Principles of Ethics of the American Medical Association demonstrated the medical profession's recognition of the confidential nature of the physician-patient relationship.

Also, in Alabama the state licensing board allows for suspension or revocation of a physician's license for betraying professional secrets. The court said that that established a legal duty of confidentiality on the part of the physician and gave a cause of action when confidentiality was breached.

The Court also discussed limitations on the patient's right of privacy. It stated that there may be situations where disclosure of confidential information without the patient's consent is justified because of some overwhelming public interest, or for the needs of the patient. However, the court did not apply this concept to the case before it.

One of the interesting features of this case is that the court did not describe the nature of the confidential information that was disclosed to the employer. Perhaps if the court had felt that the physician's disclosure was appropriate, it would similarly have disclosed.

In the recent case of <u>Anderson v. Strong Memorial Hospital</u>,[67] the patient claimed invasion of privacy and breach of the a physician-patient privilege when a newspaper photographed patients involved in an AIDS treatment and research project. The patients sued the hospital, the physician and the nurse. The patient was getting an examination in the hospital's infection disease unit. In obtaining the patient's permission for the photograph to take place the patient alleged that the physician and nurse told him the photograph would be in silhouette and he would not be recognizable. The photograph was published on the front page of the paper next to an article about AIDS research. The patient claimed that the photograph was identifiable as him and it implied that he had AIDS because of the placement of the picture next to the article. At the trial level the patient's invasion of privacy claim was dismissed because he failed the statutory test of showing that his picture had been taken without his permission for purposes of commercial exploitation. However the patient's claim of a breach of the physician-patient privilege was allowed. The patient alleged that the breach occurred in allowing the reporter in the examining room and allowing the publication of the photograph.

On appeal, the court allowed the patient's claim.
It reasoned that the physician-patient privilege statute
and related rules and regulations prohibit unauthorized
disclosure of patient information. The physician-patient
relationship carries with it an implied covenant of trust
and confidence, which is necessary to encourage people to
seek medical treatment. The court was concerned that
people who thought they had AIDS or related diseased would
be deterred from seeking health care because of the lack of
confidentiality. Thus as described by the patient, the
allegations constituted a breach of the physician-patient
relationship.

There are pending a number of cases where AIDS
patients are suing hospitals and physicians alleging
invasion of privacy, assault and battery, violation of
civil rights, and other injuries arising from their being
tested without their knowledge or consent, or from
unauthorized disclosure of their diagnosis. The CDC
guidelines[68], American Hospital Association[69] and
others promote testing with informed consent. It is too
soon to tell exactly how this litigation will turn out.

IV. Liability of Infection Control Practitioners

In examining the potential professional liability
of an infection control practitioner it is important to
understand what the basic components of such an action
are. Professional liability or malpractice actions
typically are raised as torts. The usual argument is that
the practitioner's actions were negligent under the
circumstances. In a negligence action the patient
plaintiff must first establish that the practitioner had a
duty to the patient. Secondly he/she must show that there
was a breach of that duty. A breach may occur in one of
two ways, either by an omission of an action that the
practitioner should have taken or commission of performance
of an act improperly or commission of the wrong act. The
patient must show a causally connected injury. That is, the
injury that occurred must be because of the breach of the
practitioner's duty. A breach is measured with reference
to the standards set by other infection control
practitioners. Generally, this is a national standard
rather than one based on the practice of other infection
control practitioners in the community.

Perhaps the most interesting question in looking
at these elements is whether the infection control
practitioner has a duty to the patient. Without
establishing that such a duty exists, the patient would be
unable to demonstrate that the practitioner was liable to
the patient for any harm that the patient had suffered. If
the infection control practitioner is also functioning as
the patient's attending physician or nurse, then from that
relationship, it would be clear that there is a
physician-patient or nurse-patient relationship from which
the physician or nurse has a duty of care to the patient.
This would include his/her infection control
recommendations as they relate to that particular patient.
If however, the infection control practitioner is called in

as a consultant to the attending physician, one can question whether the practitioner is merely offering advice to the attending about the attending's patient or whether the nature of the consultation itself is enough to create a physician-patient relationship and a duty to that patient.

Looking beyond responsibility to a specific patient, if an infection control practitioner is responsible for recommending or implementing a particular infection control program that affects a number of patients one may argue that those responsibilities create a duty to the patient, even though the practitioner does not know the patients who will be involved in the infection control program and does not treat them directly.

A more likely argument however, is that the hospital owed a duty of care to the patient, as established by some of the statutory, Joint Commission, and other principles discussed above. Part of the hospital's responsibility to the patient includes provision of an infection control program. The hospital discharges this responsibility through physicians, nurses, and others. These individuals act as agents or employees of the hospital. Their negligence in discharging the infection control function would be imputed to the hospital on an agency or _respondeat superior_ theory. Following the reasoning in the _Darling_[70] case, the hospital might be found directly liable. For example, if it failed to fund an infection control program or ignored statutory requirements for disposal of infectious waste, these could be treated as omissions by the hospital itself.

If a patient succeeded in a claim against the hospital based on the actions of its agents or employees, the hospital may seek to recover its losses in an action against the infection control practitioner.[71] The hospital may also discipline or terminate the employee for failing to fulfill his/her job responsibilities, seek to restrict or terminate the practitioner's medical staff privileges, or take other remedial action.

An additional source of potential liability is between a health care worker and the infection control practitioner. The health care worker might argue that the failure of the infection control practitioner to implement a sound infection control program in certain areas or with regard to the particular health care worker caused injury to the health care worker. Again, the question is whether the practitioner owes a duty to the health care worker or whether the health care worker's claim is really against the hospital. Creation of a physician or nurse-patient relationship, because the infection control practitioner is treating the health care worker, would establish a duty of care to the patient for breach of which the patient could sue. This would apply even though the treatment is made available by the hospital as an employee benefit.

Typical malpractice actions do not target infection control practitioners.[72] For example, in the case _Smith v. Curran_,[73] an orthopedist was sued when osteomyelitis developed after the patient's broken leg was operated on and set. The infection was cultured and found to be staphylococcus. The plaintiff used a res ipsa loquitur theory, arguing that the occurrence of the infection in and of itself demonstrated malpractice. The only two witnesses at the trial were the patient plaintiff and the defendant physician. The doctor received a directed verdict which was upheld on appeal.

The physician argued that the patient received the standard of care and that he did not know the source of the infection. He also argued that infections are a recognized complication of acceptable medical practice. The patient failed to offer evidence of the breach of duty by the physician to the patient and offered no evidence of the standard of care to contradict that offered by the defendant physician. The court stated that experts were needed to testify regarding the source and cause of the infection and that the plaintiff's theory was inadequate to demonstrate negligence. In other words, a bad result alone does not constitute negligence.

In contrast, an infection control program and the hospital's responsibility for it were addressed in Kapuschinsky v U.S.[74]. This was a malpractice action brought under the Federal Tort Claims Act. It involved a premature newborn who developed jaundice and other problems while she was in the premature nursery. The patient was diagnosed as having osteomyelitis of the right and left femurs, pelvic girdle, and the right humerus. The source of the infection was diagnosed as staphylococcus aureus. The injuries were permanent and resulted, in among other things, a leg length discrepancy of two inches.

Once the patient was diagnosed the nursery staff also were cultured. A nurse was found to be positive. She had not been tested or received a physical examination or any other health screening prior to starting work in the newborn nursery. She had previously worked elsewhere where she had come into contact with sick patients. There was testimony that the nurse handled the baby and had the same infection as the baby. This created the inference that the likely causation for the baby's infection was transmission from the nurse.

The plaintiff argued that the defendant was negligent under a national malpractice standard, relying in part on the Joint Commission.[75] The defendant urged that a community standard applied and offered testimony that the community practice did not include routine screening or testing of employees.

The physician in charge of the nursery testified in support of the community standard. He said that it was better not to know the results of testing, especially for asymptomatic carriers who were nurses or other employees. He was concerned that he would not hire known carriers and so would not have enough staff to run the nursery. The physician also testified that premature infants were at great risk of infection.

The court rejected the community standard. The court found that the risk of infection to the baby was foreseeable. The duty to the patient was breached in two respects: 1) that the hospital permitted contact by an inexperienced nurse with a baby who was very susceptible to infection and 2) that it permitted contact by the nurse without requiring a prior physical examination or appropriate laboratory tests. Either breach was considered sufficient to allow the malpractice action to proceed.

There is an increased public recognition of the role of infection control practitioners and the importance of an infection control program in responding to the AIDS crisis. This awareness may increase litigation against infection control practitioners and hospitals where infection and other complications occur, both to patients and health care workers.

1. This discussion represents the views of the author. It is not offered as legal advice.

2. Pa. Stat. Ann. tit. 35 §521 et seq.

3. Id. at §521.4(a).

4. Id. at §521.4(b).

5. Id. at §521.4(d).

6. Id. at §521.2.

7. Id. at §521.11.

8. Id. at §521.11(a.2)

9. Id.

10. Id. at §521.8.

11. Id. at §521.13.

12. Id.

13. Id.

14. Id. at §521.13(b).

15. Id. at §521.14(a).

16. Id.

17. These sections are good examples of the incomplete attention that may be given to another medico-legal question when the emphasis is on a different topic. Thus, with the focus on prevention of venereal disease, consent for blood tests and consent of minors is briefly addressed. Here, the statute references testing in the absence of "dissent"; other state legislation defines consent more completely. Other

provisions which affect minors use a different age at which they can act independent of their parents, or describe other health care needs e.g. pregnancy, where a minor's consent is sufficient to authorize treatment. It would be an interesting, a possibly even useful exercise to collect and compare these provisions for purpose, consistency, and clarity.

18.   Id. at §521.15.

19.   West's Ann. Cal. Health & Safety Code §429.36.

20.   228 Cal. Rptr. 700, 183 Cal. App. 3d 1272 (1986).

21.   West's Fla. Stat. Ann. §232.032.

22.   Joint Commission for the Accreditation of Health Care Organizations, Infection Control, Accreditation Manual for Hospital, pp. 75-76, 1988.

23.   See, for example, The Hospital Association of Pennsylvania, AIDS Guidelines for Hospitals, Feb., 1988, AIDS Task Group of the American Academy of Hospital Attorneys, AIDS and the Law: Responding to the Special Concerns of Hospitals, Spring, 1988.

24.   211 N.E. 2d 253, 33 Ill. 2d 326 (1965).

25.   This requirement should not be confused with the well-known "Good Samaritan" statute. This statute exists in every state in similar format. It may cover physicians, nurses, emergency medical technicians, and others. It does not compel emergency treatment. Rather, it immunizes the "treater" from liability for providing gratuitous, emergency treatment in good faith without wilful or gross negligence. See, for example, Pa. Stat. Ann. tit. 42 §8331.

26.   42 USCA §§291, 291c(e) (Hill Burton Act).

27.   42 UCSA §1395dd.

28.   28 Pa. Code §105.11.

29.   29 USCA §794 (Section 504 of the Rehabilitation Act of 1973).

30.   Supra, note 22.

31.   37 Morbidity & Mortality Weekly Rep. No. 24, (June, 24, 1988).

32.   OSHA Instruction CPL 2-2.44A.

33.   Ethical Issues Involved in the Growing AIDS Crisis, Report of the Council on Ethical and Judicial Affairs, American Medical Association, 1987.

34.   Id.

35.   42 USCA §300a-7.

36. A recent law review article cites forty-four states. Durham, Wood, & Candie, Accommodation of Conscientious Objection to Abortion: A case study of the Nursing Profession, B.Y.U.L. Rev. 253, 308 (1982).

37. Ill. Ann. Stat. Ch. 111 1/2 §5302.

38. Iowa Code Ann. §146.1; Ky.Rev. Stat. U311.800(5)(b)-(c); Pa. Stat. Ann. tit. 43 §955.2(b)(2).

39. 199 N.J. Super. 18, 488 A.2d 229 (1985).

40. Id. at 26, 488 A.2d at 233 (citing American Nurse's Association, Code for Nurses with Interpretive Statements, 1.4 at 5 (1981)).

41. This was adopted by the Joint Commission, at the April, 1987 meeting of its Board of Governors.

42. Query whether this was the real problem in the Warthen case.

43. 29 USCA §651 et seq.

44. No. 93A02-8707-EX-278 (Ind. Ct. App., 4th Dist., Apr. 4, 1988).

45. Supra, note 43.

46. Munic. Code Ch. III, Art. 5.8, §45.82(A)(1) (1985).

47. Phila. Exec. Order No. 4-86 (April 15, 1986).

48. Chalk v. U.S. District Court, 832 F.2d. 1158 (9th Cir., 1987); Doe v. Senacola & Sons Excavating, No. 86-320828N2, Mich. Cir. Ct., Oakland (Apr. 8, 1987), Doe V. Orange Co. Department of Educ., F.Supp. (USDC, C.D. Cal. No. CV87-516903). California Fair Employment and Housing Comm. v. Raytheon Co., No. FEP 83-84 (Feb. 5, 1987).

49. 29 USCA §793.

50. Pennsylvania Medical Society Policy on AIDS, Oct. 25, 1987.

51. Note that, workers' compensation depends on the individual's employment status. Thus it would not be available to a physician with only a medical staff appointment to the hospital, since that does not create an employer-employee relationship. The physician's remedy would be an action against the hospital, most likely based on negligence. As in most negligence actions, the physician's conduct in contributing to or avoiding the harm may be an issue.

52. No. 1-87-1447 WC (Ill. App. Ct., 1st Dist. May 15, 1988).

53. 185 Ga. App. 771, 365 S.E.2d 549 (1988).

54. <u>Supra</u>, note 22 at xi.

55. 28 Pa. Code §103.22.

56. <u>Id</u>. at §103.22(b).

57. <u>Id</u>. at §103.22(b)(5).

58. <u>Supra</u>, note 22 at xii.

59. <u>Supra</u>, note 22.

60. <u>Supra</u>, note 45.

61. <u>Supra</u>, note 45 at §103.22(b)(11).

62. The hospital is required to provide for the physical isolation of patients. See, for example, 28 Pa. Code §146.2 and note 6, <u>supra</u>.

63. Del.C. tit. 14 §131.

64. M.G.L.A. Ch. 76 §15.

65. Ill. Rev. Stat. Ch. 111 1/2 p. 7506 §6.

66. 291 Ala. 701 (1973).

67. No. 4541/88 (N.Y. Sup. Ct. July 27, 1988).

68. <u>Supra</u>, note 31.

69. American Hospital Association, <u>AIDS/HIV Infection Policy</u>: <u>Ensuring a Safe Hospital Environment</u>, Report and Recommendations of the Special Committee on AIDS/HIV Infection Policy (Nov. 1987).

70. <u>Supra</u>, note 24.

71. There is no incentive to sue the infection control practitioner if all are part of the same insurance program.

72. This may be due in part to lack of documentation in the patient's chart that an infection control practitioner was involved in the patient's treatment, or that treatment was (or should have been) subject to an infection control protocol.

73. Colo. Ct. App. 472 p.20 769 (1970).

74. 248 F.Supp. 732 (USDC,D So. Carolina, Charleston Div. 1966).

75. <u>Supra</u>, note 22.

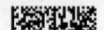